CÉ

D0893501

THE MINERS OF
NOTTINGHAMSHIRE
1914-1944

1. F. B. Varley, MP, A. J. Cook and Herbert Booth at a meeting on Hucknall Recreation Ground during the 1926 lockouts

THE MINERS OF NOTTINGHAMSHIRE
1914-1944

A HISTORY OF THE NOTTINGHAMSHIRE MINERS' UNIONS

ALAN R. GRIFFIN

REPRINTS OF ECONOMIC CLASSICS

AUGUSTUS M. KELLEY • PUBLISHERS
NEW YORK • 1966

PRINTED IN GREAT BRITAIN
in 10 *point Times Roman*
BY C. TINLING & CO. LTD
LIVERPOOL, LONDON AND PRESCOT

TO IAN WINTERBOTTOM

FOREWORD

The period covered by this volume saw momentous changes in the structure of trade unionism in the mining industry.

In 1914 every coalfield had its own District Union. These district unions were bound together into a loose Federation (the Miners Federation of Great Britain). During the war years the locus of power shifted perceptibly from the district associations to the national Federation. This shift in the locus of power was welcomed by what one may call the Federation's left wing who wished to convert the loose Federal structure of the union into a strong centrally controlled union. The internal struggles which took place in the Federation during the 1920's were between those who accepted the left wing view and others, notably Nottingham's George Alfred Spencer, who believed that power should continue to reside with the separate district associations.

This argument was based partly upon ideological and partly upon economic grounds. The left wing tended to regard trade unions as instruments for securing the common ownership and control of the instruments of production. On the other hand, Spencer and men who thought with him, regarded trade unions primarily as organizations whose duty it was to secure for their members the maximum economic benefits within the framework of the existing Capitalist Society.

The economic basis of the argument between left and right wings of the Federation is to be found in the wide variations in profitability of the individual coalfields. With the possible exception of Leicestershire and South Derbyshire, Nottinghamshire was much the most profitable district. It, therefore, stood to gain the most by remaining independent. Any pooling of wages on a national basis would be bound to lead, in the circumstances of the time, to a levelling down of wages in the better paid districts and a levelling up in the less well paid districts. Looking at this from the point of view of sectional interests, therefore, Spencer could say that a National organization with a National wages ascertainment would be wrong.

The present volume describes the events which followed from this difference of view between the two wings of the Federation. It presents an unbiased picture of the struggles in the Nottinghamshire coalfield in the 1920's and 1930's culminating in the formation of the Nottinghamshire Miners' Federated Union in 1937. It then traces the history of this organization up to the year 1944 when the decision to form a National Union of Mineworkers was taken; and it will be noted that the Nottinghamshire Area of the NUM has staunchly supported the national union and the nationalized industry since its inception and has loudly voiced its protests at the many suggestions made by the Tories for decentralization.

We believe that such an unbiased picture as is here presented is a necessary antidote to the partisan stories which have found a ready circulation from time to time.

J. T. TIGHE *President*
A. MARTIN *General Secretary*
L. CLARKE *Financial Secretary*
W. L. ELLIS *Agent*
W. BAKER *Agent*

National Union of Mineworkers
(Notts. Area)

PREFACE

During the period covered by this volume, Nottinghamshire has played an important part in the affairs of mining trade unionism. For this, and other reasons, it has been necessary to go over much of the ground already covered by Mr Page Arnot in his history of the Miners' Federation of Great Britain. I have tried to avoid treading too heavily on Mr Arnot's heels and in consequence my account of national matters and proceedings is, at times, sketchy. I am particularly unhappy about the chapter on the General Strike. Had it been possible, I would have omitted this altogether since the story has been told by so many different people in so many different ways. Since it was necessary for completeness to deal with the strike, however, I have condensed the story as much as possible at the risk of being accused of writing potted history.

In writing this volume, I have drawn very heavily on the Minute Books of the Miners' Federation of Great Britain which contain verbatim accounts of conferences, Executive Committee minutes, and documents of various kinds. The books and articles consulted are listed in a bibliography at the end of the volume. For this period it was not necessary to rely so much on the files of the local press but I have, nevertheless, made fairly extensive references to them.

I have received information and advice from a great many different people, and in particular from Mr Herbert Booth, with whom I have corresponded regularly for some years. Other people who have helped me in various ways include Miss L. Walvin, Mr A. A. Booth, Mr Frank Dennett, Mr W. L. Ellis, Mr Bernard Taylor, MP, Ald. W. Bayliss, JP, the Rt Hon Emmanuel Shinwell, MP, Mr Owen Ford, Mr A. C. Overfield, Mr O. B. Lewis, Dr J. E. Williams of Sheffield University, the late Mr J. M. R. Watson of Nottingham University, the late Mr Ernest Mellors and the late Mr G. A. Spencer. To all of these, and to the many miners and ex-miners to whom I have spoken, I express my thanks. I am grateful, too, to Mr George Spencer's family for letting me have various books and papers belonging to their father.

Dr A. W. Coates of the University of Nottingham and Mr Herbert Booth have been good enough to read my typescript and to make various suggestions for its improvement. Some of these I have adopted. For the errors, omissions, and general inadequacies of the volume, however, I claim sole credit.

Many of the footnotes which adorned my typescript have been taken out in an attempt to make the book a little less tedious to read.

The rest have been relegated to the end of the book; and my original draft is available for anyone who would like to make detailed references to my sources.

A.R.G.

Dept. of Economic History,
University of Nottingham,
December 1961.

CONTENTS

NOTE :

The term Council, used without qualification, refers to the Council of the Notts. Miners' Association or the Notts. Miners' Federated Union. Council consists of one delegate from each Branch, and is the Union's governing body.

ILLUSTRATIONS

INTRODUCTION

Our earlier volume traced the history of the Nottinghamshire Miners' Association down to the beginning of the first World War. By 1914, the Association had settled down into a humdrum existence. Membership was expanding gradually as the labour force grew, funds were steadily accumulating and relations with the employers were good. No one at that time could have foreseen the difficulties which the union was to experience in the years ahead: no one, for that matter, could have foreseen the economic difficulties which were to face the mining industry in the post-war years; and the two were very closely connected. Within a matter of twelve years the Association was to lose all its money, most of its members, many of its leaders, and its good relations with the employers.

The history of the Nottinghamshire Miners' Association in our period is of more than local interest. The success of the 'Breakaway' non-political union formed in 1926 under the leadership of George Alfred Spencer, caused a great deal of concern to the leaders of the Labour Movement. For the first time in British trade union history, a serious attempt was made to build up a non-political trade union organization in opposition to the Trades Union Congress. Offshoots of the 'Miners' Industrial Union' were formed in all the principal coalfields, and a 'Non-political Trade Union Movement' with Havelock Wilson and George Spencer as its leaders, was founded.

As it happens, the TUC, in the period following the General Strike, adopted a policy of peaceful co-operation with the employers. Had the militant attitude of the early 1920s not been abandoned, employers as a class would doubtless have thrown their weight behind Wilson and Spencer; but in the event this was unnecessary. The TUC of its own volition abandoned the idea of using industrial disputes for political purposes; and instead pursued a policy of peace within industry. By the time the 'Non-political Trade Union Movement' came into existence, therefore, it was already an anachronism.

In this volume, an attempt is made to show that the Spencer 'breakaway' was the logical outcome of a fundamental disagreement on the purposes of trade unions; and on the strategies and tactics which they should employ in particular situations. For Spencer, the purpose of a trade union was to co-operate with the employers to make their undertakings profitable and then to see that the workers received their 'fair share' of the proceeds. It follows that since undertakings (and, in the mining industry, districts) differ in their ability to pay, wage negotiations should be undertaken locally.

B

Spencer's opponents, on the other hand, wished to see an end to private ownership; and they felt that wage-rates should be negotiated nationally so as to reduce, or even eliminate, district differentials.

Spencer's views were undoubtedly coloured by his awareness of Nottinghamshire's favoured economic position. Throughout the whole of our period wage-rates were higher in Nottinghamshire than in any other county owing to the superior productivity of her mines.

Nottingham has been favoured by good geological conditions. There are many workable seams of coal within a few hundred yards of the surface in most parts of the county. Most of these seams are of good quality coal; are reasonably free from faults, and are fairly easily worked. The prosperity of the Leen Valley (which was, in the great days of the mining industry, the most prosperous mining district in Britain) was founded upon the Top Hard seam which outcrops at Wollaton. This seam is now being intensively worked in the Northern part of the county.

However, geological conditions alone do not account for Nottinghamshire's leading position. Yorkshire, whose geological conditions are probably rather better than Nottingham's, has a lower output per man shift. It may be that the 'Butty' system, which operated in Nottinghamshire but not in Yorkshire, has resulted in a higher tempo of work in the former county. Technological factors must also be taken into account. In the days when inadequate transport facilities restricted the size of the market, the Nottinghamshire coalowners were spurred by the intensity of competition into investing in capital projects of various kinds. Rails, boring rods, efficient drainage systems and pumps were all pioneered in this county. In our period such firms as the Bolsover Company; Butterley; Barber Walker and B. A. Collieries pursued progressive investment policies. By 1938, some 70 per cent of the county's output was mechanically cut (against a national average of 59·5) and 82 per cent was mechanically conveyed (against a national average of 54·2).

Despite the relatively high level of wages in the county, the labour costs per unit of output are low. In a typical year, 1949, wages costs, at 23s 2d per ton, were the third lowest in the country, and the profit at 9s 1d per ton, was the highest. The margin of profit in Nottinghamshire provides elbow room in which wage negotiations can take place: whereas in districts with a slender profit margin the struggle for improved conditions (or even for the maintenance of existing standards) tends to be bitter. This accounts in part for the satisfactory labour relations enjoyed by Nottinghamshire in comparison with such districts as South Wales and Scotland.

It will be remembered from our first volume that, in 1893, some of the Nottinghamshire coalowners were reluctant to join in the national lockout since they were still able to make a profit despite the fall in coal prices; and that, half-way through the dispute, they re-opened their pits at the old rate of wages thus causing a 'split' in the coalowners' association. This event provides an interesting parallel with the Spencer 'breakaway' of 1926. The basic cause of the 'breakaway' was the same in both cases: the superior profitability of the Nottinghamshire pits.

Fortunately, in 1893 the Miners' Federation of Great Britain allowed men to return to work at any colliery where the employers were willing to pay the pre-stoppage wage. Had they then adopted the attitude which prevailed in 1926—that all districts must remain out until the dispute was settled—it is possible that the NMA would have broken away from the Federation at that time.

The outcome of the 1893 lockout was much more favourable to the Federation than that of 1926. In the earlier dispute, the men eventually returned to work at their old rates of wages, whereas in the later dispute severe reductions were suffered and hours of work were increased. Why was this? The chief reason is that in the earlier period, despite a temporary recession, the long-term demand for coal was rising whilst in the later period the demand was falling. Further, in 1893, the public sympathized with the miners and gave a great deal of tangible support, both in cash and kind; whilst in 1926 the fear of 'red revolution' robbed the miners of middle-class sympathy. The levies raised by the men at work in 1893 also helped: there was nothing of the kind in 1926. Again, the 1926 lockout occurred at a time when the constituent Associations of the Federation were weakened by the major disputes of 1920 and 1921 and by the existence of mass unemployment which drained their funds and undermined their morale. The 1893 lockout was fought with an élan altogether absent in 1926.

The 1893 lockout strengthened the MFGB but the 1926 lockout weakened it. Following the earlier dispute, a Conciliation Board for the Federated District (i.e. all the major coalfields except Northumberland, Durham, South Wales and Scotland) was established, but following the 1926 dispute, the owners refused to discuss wages matters at all except on a local basis. Their intransigence in this—and other—respects made the nationalization of the mining industry an absolute certainty for a future Labour Government with a working majority.

By the time the Labour Party achieved power the Mineworkers' Federation of Great Britain had been converted into the National Union of Mineworkers. The 1944 Annual Conference of the MFGB at which the decision to form the NUM was agreed in principle was held in Nottingham; where, some eighteen years earlier, Spencer had raised the standard of non-political unionism. The story of these two events, and of the intervening years, is central to the study of recent trade union history, and it forms the core of the present volume.

CHAPTER 1

BUSINESS AS USUAL

1. *General*

It is appropriate that this 'nation of shopkeepers' should have adopted the slogan, 'Business as Usual' in the dark days of 1914. However, the desire to do business as usual was tempered, in the Nottinghamshire coalfield as elsewhere, by feelings of patriotism. Mr Dennis Bayley (of the Digby Colly Co.), for example, opened a fund to provide and maintain a Red Cross Ambulance Convoy; a fund to which the Nottinghamshire miners contributed some thousands of pounds. But, for the coalowners, the war meant rich pickings. Coal prices soared, whilst wage-rates limped slowly upwards (at least until 1918 when there was a marked improvement). Indeed, the widespread unrest in the coalfield was held in check only by patriotic feeling and by the regularity of employment. Wage rates, measured in real terms, fell; but average real earnings probably rose.

In the period 1909-13, the average annual earnings of miners in the country as a whole were £82. The average for 1914 was a little below this figure at £79. The average then rose to £105 in 1915; £127 in 1916; £129 in 1917 and £159 in 1918. 'Real' earnings, that is, earnings measured in terms of what money will buy, rose by about 14 per cent in 1915, remained more or less the same in 1916 and then fell to something like 89 per cent of the 1914 level in 1917 when earnings remained almost stationary despite steeply rising prices of food and other consumer goods; and then rose to something like 5 per cent over the 1914 level in 1918.

Had the pits not been making good time, then the miners and their families would have been considerably worse off than in the immediate pre-war years, however, since piece-work and day-work rates failed to keep pace with the rise in the cost-of-living. The lower paid men were, in one respect, more fortunate than the more highly paid piece-workers. The flat-rate increases introduced during the war diminished the differentials, so that the real wage-rates of the lower paid men probably rose (or, at least, fell only very slightly), whilst the real wage-rates of the piece-workers show a marked reduction. Thus, in the Erewash Valley and Derbyshire District, the average wage per man-shift for day paid coal getters rose from 6s 10d, in June 1914, to 13s 9d, in November 1918; but the average wage per man-shift for piece-work coal getters rose from 9s 11d to 17s 7d.

In the first case, the November 1918 wage is practically double the June 1914 wage; but in the second case, the increase is one of 77 per cent.

2. *The Nottingamshire Miner, a Model Patriot*

The Nottinghamshire miner was a model patriot.

Every week he subscribed one penny to the Red Cross Ambulance Convoy fund in addition to paying special levies (e.g. to provide the old-age pensioner members of the Association with an extra half-crown a week) and his ordinary union contribution.

Early in the war, Council resolved that:

'We warmly appreciate the efforts being made by our members and the general public to relieve distress caused by the war; our only regret being that organized labour has not a larger representation on some of the local distress committees.'

The Association also showed its disapproval of the attitude adopted by the ILP Anti-war party in the Miners' Federation of Great Britain. This led to a suggestion that the NMA should withdraw from the MFGB Political Fund and form one of its own:

'. . . so that the NMA, may itself become a political unit, frame its own policy and Rules, and completely control its own fund and political interests both Imperially and locally.'[1]

This move to break away from the MFGB politically was precipitated by the withdrawal of the Labour Whip from J. G. Hancock on account of his continued membership of the Liberal Party. Since Hancock was no longer nominally a Labour Member, he lost the £100 a year Parliamentary Allowance (plus election expenses) which Labour miners' MPs drew from the MFGB Political Fund.

Hancock was supported in his attempt to disaffiliate the NMA from the MFGB politically by Spencer, who later became a Labour MP himself. Indeed, the wording of the resolution quoted in our previous paragraph bears Spencer's hallmark. However, a campaign protesting against the proposed move was launched by Herbert Booth, recently returned from the Central Labour College, and W. Askew, of Newstead. They addressed meetings up and down the county and they issued at their own expense 30,000 copies of a leaflet exposing the attempt to secede. They were helped by a ruling from the Registrar of Friendly Societies that a ballot vote would need to be held before the proposed new political fund could be established[2] and they insisted that a ballot vote should be held.[3] As a result of this ballot, the attempted breakaway was squashed. This episode is not, however,

[1] See Footnotes at end of book.

without significance. Indeed, it can be regarded as the opening move in the struggle for a 'non-political' union and, at the same time, as the final attempt of the out-moded Lib.-Labs. to keep the union tied to the tail of the Liberal Party.

Two years later, on June 27, 1917, Council resolved that:

'Seeing that the Labour Conference by a large majority passed a resolution to help the Government to prosecute the war vigorously and to a successful issue we deprecate and strongly protest against the attitude taken up by the [anti-war] minority of the Labour Party, including the President of the Miners' Federation of Great Britain, Mr Robert Smillie and Messrs Snowden, Macdonald, Jowett and others.'

Again, a special meeting of Council, held on August 18, 1918, expressed its disapproval of the proposed International Socialist Conference at Stockholm, and instructed its delegates to vote against the proposal.

The union also agreed to assist the recruiting authorities to 'comb out' men for military service. In the early days of the war, many miners had volunteered for military service, and in consequence, a large number of men from other trades entered the mines. When the authorities extended the principle of compulsory enlistment to the mining industry, the union in company with the other district associations, insisted that these 'outsiders' should be taken first. But during the later stages of the war, the Association agreed to operate a scheme whereby the quota of recruits demanded by the authorities should be chosen by lot. Thus had the union become a recruiting sergeant.

Its special position enabled the Association to take up with the authorities complaints made by called-up members. Thus, on November 28, 1917, Council resolved that:—

'The attention of the Federation Executive be drawn to the brutal manner in which Army Officers are treating Privates in some cases.'

Complaints were also made to the military authorities about inadequacies in the quantity and quality of the food being supplied to soldiers.

However, the patriotism of the association did not extend to allowing the owners to lengthen their working day. Despite representations made by the Government, the union firmly refused to entertain any suspension of the Eight-Hour Act. The union did, however, advocate the shortening of holidays, and it supported the campaign to coerce persistent absentees into attending their work more regularly. Union representatives took part in the work of the absentee committees which were set up at colliery and district level. However, these committees were largely ineffectual if reports are to be believed.

Union representatives also served on local food committees. Towards the end of the war, and just after, Council made repeated representations to Mr Marsden Smedley, the Food Controller, in order to ensure a more equitable distribution of food. At a meeting on February 2, 1919, for example, the following motion was adopted:

> 'That the Council strongly protest against the present method of distributing the essential foods and against the present composition of many of the local food committees of the County, it further desires to call attention to the growing unrest of the unions and other workers in the County owing to unequal distribution of food and declares that unless some immediate steps were taken to remedy the growing grievances on the lines of rationing, any dislocation of trade or stoppage of industry from the above cause will be due either to the inactivity or inability of the responsible authorities to enforce that end, that a committee consisting of the President, Vice-President, with officials, seek an interview with Mr Marsden Smedley to bring the question before him.'

Finally, when the war was over, the union co-operated with the authorities responsible for demobilization to ensure the smooth absorption of members returning from the forces into the industry.

3. *The Federated Area*
We saw in the first Volume that, from the 1890s, the miners' unions of the central coalfields acted in concert in the matter of general wage adjustments. Owners' and workmen's representatives met periodically to consider suggested variations in wage rates; and, in case of a disagreement between the two sides, an independent chairman could be called in to give his casting vote.

The Conciliation Board Agreement provided for a maximum addition to the 1888 basis of 65 per cent. Having reached their maximum, therefore, the miners could not apply for any further increases until March 31, 1915, when the Agreement was due for renewal.

However, during the winter of 1914-15, the prices of foodstuffs and other consumer goods were rising sharply. Speculators were felt to be making a good thing out of the war at the expense of miners and other workers. On February 27, 1915, the Nottinghamshire Miners' Association Council resolved:

> 'That we enter our strong protest against the cornering of

essential commodities, and the inflated prices that are being charged, especially under present trying conditions, and urge the Federation and Government to protect the People so far as possible against these exploiters.'

At its previous meeting, Council had been discussing a demand for a general advance in wage rates, and had instructed its representatives on the Conciliation Board to 'press for more favourable conditions so far as minimum, maximum, future advances in wages, and surface workers are concerned.' Negotiations were going forward on a national basis for a 20 per cent advance in earnings, but the owners' organization insisted that the question should be dealt with separately in each district. On May 26, the Secretary of the NMA, Mr C. Bunfield, reported that, in the Federated Area, a War Bonus of 15½ per cent on current earnings (not on standard) had been conceded. In accepting his report, Council resolved 'That we express our very high appreciation of the sound wisdom and admirable loyalty shown by our members in continuing to work during the prolonged wages negotiations and notwithstanding the cessation [of work] in other districts.'

The MFGB, at its 1913 Annual Conference, held at Scarborough, had decided to give notice to terminate all Conciliation Board Agreements together, with a view to placing all the Districts on the same basis. This notice was given, and subsequently a new arrangement was agreed on in the Federated District. It was decided to merge 50 per cent of the advance on the '1888' basis into the standard rate, thereby forming a new basis (called the December 1911 Basis). All future percentage variations in wage rates were to be calculated on this new basis.

During the war, the Conciliation Board for the Federated District continued to meet at regular intervals to discuss wage claims. One such meeting, with the independent chairman, Lord Coleridge, in the Chair, was held on Wednesday, March 8, 1916. In putting the men's case, Mr Stephen Walsh, MP, Chairman of the Workmen's side, pointed out that:

'Since December 1911, the date from which the new basis rates are calculated, wages, including the (war) bonus, have increased by 32 per cent, selling prices by 68 per cent. In the statement previously made by the employers, it was admitted that, to pay a 5 per cent advance on the old basis, equalling 3⅓ on the new, to those governed by the decisions of the Board, 3d per ton was required; and it follows, even upon the employers' own contention, to which we do not bind ourselves, that to pay the 32 per cent on the new basis, a sum of 2s 4¾d, only would be necessary. As the employers have obtained a clear increase in selling prices from 7s 7·26d, to 12s 9·85d, or 5s 2½d per ton, thus leaving a clear margin of 2s 9¾d per ton after meeting all wage claims, we submit that these facts alone more than warrant the granting of our application.'

In his reply, Mr F. J. Jones, for the owners, urged that wage increases were not in the national interest. He pointed to statements made on behalf of the Government calling for the restriction of 'any further advances of wages . . . to the adjustment of local conditions.' He also said that:

'The effect of the large absenteeism from work, we think abundantly proves that the wages paid to miners are more than sufficient, or they would not be able to absent themselves from work, and so, materially reduce the wages that they might receive by a more regular attendance.'

He also contended, of course, that the industry could not afford to pay the increase demanded, which was one of 5 per cent on the 1911 basis.

After listening to the arguments, Lord Coleridge gave his casting vote in favour of the advance.[4] A further advance of 3⅓ per cent was conceded by the owners, without reference to the Independent Chairman, in June 1916. The leaders of the miners' unions realized all too well that, in the battle between rising prices and rising wage rates, wage rates were coming off second best and efforts were made to spur the Government into stemming the rise in the cost-of-living, but without success. Faced with the phenomenon of constantly rising prices, the men continued to press for wage increases. In February 1917, an increase in the War Bonus, equivalent to 4·2 per cent on current earnings was secured. Then, at its meeting on April 30, 1917, Council proposed that the MFGB should seek a further substantial increase in the bonus. However, the workmen's side on the Conciliation Board had undertaken to make no further applications within the period ending August 31, 1917 and the Nottingham suggestion could not, therefore, be entertained.

In South Wales, Durham and Scotland the owners had actually demanded reductions in wages in the Winter of 1916-17 owing to a recession in the export trade coupled with lowered production consequent upon the enlistment for military service of large numbers of miners; whilst in Northumberland, a reduction of 11 per cent had actually been forced on the men in December 1916, despite rising food prices; and a further reduction was asked for early in 1917. But the Government had now taken a hand in the running of the industry. The South Wales coalfield was placed under a measure of State control on December 1, 1916; and this control was extended to the other coalfields on March 1, 1917.

The Coal Controller refused to countenance the owners' demands for reductions in wages; and on September 17, 1917, he granted a universal flat-rate increase of 1s 6d for men and 9d for boys. This was called the 'War Wage.'

In 1918, the cost-of-living rose faster than ever. Branches were pressing for a further general increase and, on May 6, 1918, Council resolved to urge the MFGB to propose that the flat rate War Wage should be increased from 1s 6d to 4s 0d a shift (9d to 2s 0d for boys). However, the claim made by the MFGB was for an increase of 1s 6d per day for men and 9d per day for boys, and this was conceded, very reluctantly, on June 30.

On August 27, a further resolution was put forward by Council, demanding an increase of 6s 0d per day for men and 3s 0d for boys. This resolution was reiterated at a further meeting on New Year's Day, 1919, when it was suggested that, failing an early settlement, a national strike ballot should be taken.

In our next Chapter, we shall examine the fate of this claim, but in the meantime, perhaps we should complete our survey of the history of the Federated District.

The Federated District first came into existence in 1888, when the foundations of the Miners' Federation of Great Britain were laid. Its constituent areas, Yorkshire, Derbyshire, Nottinghamshire, the Midland Federation, Lancashire, Cheshire and North Wales, produced the bulk of their output for the home market. They were, therefore, insulated to some extent from the vagaries of international trade which kept Northumberland, Durham and South Wales tied to sliding scales. For a long time, these latter 'outside' districts refused to join the Miners' Federation of Great Britain, and during the great lockout of 1893, they continued to produce coal. Eventually, they joined the MFGB, but still kept their separate Conciliation Boards. The original members of the MFGB, who formed the Federated District, retained their old organization, which became a federation within a federation, so to speak.

The Conciliation Board for the Federated Area, consisting of representatives of the two sides, held quarterly meetings where general issues affecting the District were discussed, and they also had special meetings from time to time to discuss applications for alterations in wage rates. The men's representatives on the Conciliation Board held regular meetings to formulate policy (especially in relation to wage claims). Moreover, annual and special conferences were also held to which only the district unions in the Federated Area were invited to send delegates.

At the 1917 Special Conference, the Chairman, Mr S. Walsh, said:

'The whole idea of this Conference of this area was so far as possible to try and find out what are the points upon which there is the greatest measure of agreement amongst ourselves so that at the Annual Conference of the Miners' Federation of Great Britain we shall not go in as discordant atoms . . . in the past this area, which commands by far the greatest output in the whole of Great Britain, has gone into the Conference, I do not want to say shackled, but shorn of a great deal of its effective strength, because we have never tried to come together beforehand. There can be no harm from trying to come together on points of agreement, but every county will be left with a perfect right to take their own course on points of non-agreement.'

This point of view, that the Federated District should retain its separate identity—and its right to safeguard the interests of its constituent unions—within the wider Federation met with general approval.

But by 1918, the approval was not quite so general. One of the 'outside' districts, South Wales, had put down a motion for consideration at the Miners' Federation of Great Britain Conference, asking that 'the machinery for dealing with the general wage rate be centralized in the National Federation.' The Federated District Conference was divided on the question of whether or not to support this resolution. If the resolution were to be put into effect, then the Federated District would cease to exist as a separate entity and its Conciliation Board would be wound up.

It so happened that, at this time, the agreement to which the Conciliation Board owed its existence was due for renewal. The men's representatives thought that any renewal should be dependent on several alterations in the constitution of the Board. In particular, they did not want the Board to be shackled any longer by a provision for a maximum in the increases which it could grant, and they wanted an alteration in the method by which the average selling price was computed. Obviously, however, if the policy of South Wales were to be adopted, then there would be no point in negotiating further with the owners on these issues.

The opinion of the opponents of the South Wales resolution was well put by Mr W. Latham, of the Midland Federation, who said:

'. . . it is up to us, now, I am thinking, not to trust to what may take place in the National Conference, because even then we have been so many times put off. Why consider that? When it suits Scotland, when it suits South Wales, when it suits Northumberland

and Durham they will leave us alone, and severely, too. I am thinking about the '93 strike, and what happened then. I am hoping we shall not attach too much importance to what takes place at the National Conference. I say here and now that as an English Conciliation Board, and directing the interests of the people in this Federated Area, we ought to proceed, as we have got proposals which, if obtained, will make the future much brighter than the past for our men.'[5]

The MFGB Conference met, and the South Wales motion was adopted; but the matter was still not settled to the satisfaction of all the members of the Federated District. A further conference was called on August 23, 1918. At this conference, the principal speech in opposition to the South Wales proposal was made by Mr G. A. Spencer, of Nottingham, who said:

'I am very pleased that you have touched briefly on the history of the Conciliation Board in this part of the Federation. The Board in itself has been very effective, and that in itself is a justification for its continuance. In reviewing the Resolution that was passed at the Annual Conference [of the MFGB] I am not quite certain that the interpretation you have put upon it is the only interpretation it can bear . . . the resolution itself sets forth these terms—that owing to the limitation of profits and the limitation of prices, the Conciliation Board cannot in the future be effective, and for that reason it is essential that the function of the regulation of wages should be invested in a central authority. Now assuming for a moment that is correct, is there anyone here can say how long and to what extent a state of affairs of that character is going to continue? . . . The principle of centralization can only be desirable or effective on the assumption that after the war, when we return to normal conditions, we have secured permanently control or nationalization of the mines. . . . We have no guarantee that anything approaching that will take place. I think the whole of the evidence from the employers of this Area and the Federation is, *that they are not disposed to come together to form a National Wages Board.* They themselves will do all they possibly can to hinder that. I think, until we have secured a permanent central authority so far as the mines are concerned . . . it would be a very unwise thing indeed to disband an institution which has been so effective and so favourable *to our own particular interests.*'

Mr F. B. Varley took very much the same line. He said that, when the last application for an increase in the War Wage was made, the Federated Area could have got an increase of 2s 6d per day had it been negotiating separately with the Coal Controller, whereas the increase negotiated nationally was 1s 6d. He went on:

> 'We came here this morning to bury Caesar [i.e. the Conciliation Board of the Federated District] and we can scarcely reconcile the two; perhaps not to resuscitate Caesar, but to allow him to live in a state of coma, and in that state of coma is he going to be any good?'

He referred, as did other delegates, to the fact that the South Wales Federation was itself a party to a Conciliation Board which could not be wound up until at least nine months after the end of the war.

During these debates, several speakers showed their suspicion of the South Wales 'extremists'. It will be appreciated that, in South Wales, the syndicalist element was strong; whilst in the Midlands many of the leaders were still 'lib-lab' in outlook.[6]

One of the keenest supporters of the South Wales position in the Federated District was Mr Herbert Smith, of Yorkshire. At the London Conference, held on June 4, 1918, he said:

> 'Some of our friends say, 'What shall we do in the Conciliation Board area; cannot we get it extended?' I am not anxious to get it extended. I have had quite plenty with this. I want on the 31st July to shake off the chains. . . . As to the question of the control of mines, I am rather surprised at some of the speeches made, and I do not think we ought to say we are going to be satisfied with the control of mines, but to nationalize them—that we ought not to play with this thing. Every district should stand for nationalization, and we ought not to assume that the control of the mines is going to cease. . . .'

The attitudes taken up by Mr Spencer and Mr Herbert Smith, in 1918, pointed to the crevasse which divided the 'national action' party from those who preferred to continue to rely on district negotiations. This crevasse was to widen with the years; but in 1926, as in 1918, Mr Herbert Smith was on one side and Mr George Alfred Spencer on the other.

At the end of 1918, owners' and workmen's representatives were still meeting to discuss the possibility of resuscitating the Conciliation Board. But Caesar, even though not finally buried, had already breathed his last. The Conciliation Board for the Federated Area was as dead as the dodo. For the next decade, the industry was to be divided into two warring camps, with the owners determined to maintain the old pre-war state of anarchy, and with the Miners' Federation determined to enforce a radical change in the organization of the industry. Those who, like Mr Spencer, believed that until nationalization came along each district should get what benefits it could for its own members, found themselves in a no-man's land, subject to attacks from both sides.

4. *Local Disputes and Negotiations*

In our first volume, we saw that wages were composed of two elements: basis rates which were negotiated locally, and a percentage addition which varied according to the state of trade and which was fixed by the Conciliation Board for the Federated Area. Let us now turn our attention briefly to the local negotiations on basis rates.

So far as the Surface workers were concerned, business was very much as usual throughout the war. Council had the Surface workers' wages claim before them on August 29, 1914, when the Agent was instructed to try to reach a temporary settlement. Claims for agreements were submitted to both Owners' Associations, and in addition, negotiations were also conducted at individual pits. Thus, on April 29, 1915, it was reported to Council that a substantial increase had been secured by the Teversal surface men. It was further reported, on May 29, 1915, that an agreement had been entered into with the Erewash Valley owners (the Midland Counties Colliery Owners' Association) for certain grades, the increases to be paid from June 11. Following this, the Leen Valley owners (the Nottingham and Erewash Valley Owners' Association) were asked to come to terms in the same way. On October 30, 1915, proposals from the Leen Valley owners were considered by Council, but the Branches were unable to accept them.

On July 31, 1916, the outstanding claim for an inclusive price list for the Erewash Valley surfacemen was referred to an arbitrator, Mr W. Mackenzie, KC. At these proceedings, the men's case was presented jointly by the NMA, and the Derbyshire and Nottinghamshire Enginemen and Firemen's Union. There had been much rivalry between the two unions in the past, and the owners' side made good use of this at the Arbitration Court. Mr Piggford, for the owners, accused the NMA of:

'... poaching on Mr Rowarth's (i.e. the Enginemen and Firemen's Secretary's) preserves. He did all the hard work, and as soon as ever you found that his Association was of some value you tried to rob him of the fruits of his labour.'

This was by no means an isolated jibe; representatives of the owners returned to the charge several times. The men substantially won their case, by making comparisons with conditions in the Leen Valley. But the division between the two workmen's associations tended to widen from this date.

On December 22, 1917, Council resolved that all Enginemen and Firemen should be forced to join the NMA, and the Derbyshire Miners' Union was to be asked to take complementary action over the County border. This resolution was confirmed on January 30, 1918, after a Branch Vote had been taken on the issue. However, the leaders of the NMA were by no means as strongly opposed to the Enginemen and Firemen's Union as were their branch members, and indeed, Mr Hancock had defended the right of the rival body to lead its separate existence, at the Conciliation Board Area Conference, on September 22, 1916. Further, the Derbyshire Miners' Association refused to take the complementary action demanded of them.

During 1918, the DMA, the NMA, and the Enginemen and Firemen's Union had a series of joint meetings which resulted in the formation of a joint Board to deal with all matters of common concern.[7]

A second arbitration case was fought in August 1918, with W. H. Stoker as the Arbitrator. The Award resulting from this arbitration came into operation on September 16, 1918 and gave rates (i.e. 1911 basis rates) ranging from 6s per day for craftsmen to 4s 10d for general labourers. It applied to the Leen Valley as well as to the Erewash.

During the war, many new or revised underground price lists were negotiated: for example, at Rufford in August 1915; at Teversal in September 1917 (after the men had threatened to go on strike); at Hucknall No 1 in November 1917; at both Hucknall pits in September 1918; and at Bulwell in October 1918. Negotiations were also conducted on a district basis for new make-up rates for men working in abnormal places; and for improved pay for boys and machine coalcutters. The boys at Watnall came out on strike in June 1918, and Council condemned their precipitate action.

The coalcutters' question occupied much of the officials' time during the years 1917-18. Newcastle took a strike ballot on the issue in September 1917, and feelings ran high at several collieries. On May 27, 1918, Council decided to refer the question to arbitration, but the Erewash Valley owners subsequently made an offer for cutters on daywork, which was accepted (on November 9, 1918). This provided for the following basis rates:

Drivers and Chargemen .	.	7s 9d per shift of 8 hours.
Jibbers and Timberers .	.	7s 3d per shift of 8 hours.
Cleaners Out .	. .	6s 9d per shift of 8 hours.
Apprentices .	. . .	5s 3d per shift of 8 hours.

Percentages and War Wage were, of course, payable in addition.

The Gedling men had a series of grievances which were placed in the solicitor's hands in March 1917. On June 27, 1917, the men were authorized to hand in strike notices; on August 4th, Council decided to refer violations of the eight-hour Act to the Mines Inspector; whilst on November 28th, the men complained about the inadequacy of the facilities for seeing the manager. Finally, on January 30, 1918, the Secretary, Mr C. Bunfield, reported that most of the points in dispute had been settled.

In 1918 the men at Cinderhill complained about the treatment they were receiving at the hands of certain of the Company's officials, the chief offender being a deputy named Cheetham. The men threatened to strike if the bullying was not redressed.

At the New Hucknall and Bentinck Collieries, the men were in dispute with the management over the provision of electric lamps from early in 1917 until November 1918 when, after a short strike which cost the union over £5,000 in strike pay, the owners agreed to install efficient lamps. Subsequently Council asked that all companies should buy electric lamps.

A further series of disputes during the war years arose from the employment of non-unionists—many of them newcomers to the industry. This question was raised at Brierley Hill and at the Babbington pits in July 1917. Strike ballots taken at both places resulted in large majorities in favour of terminating contracts. Most of the non-members at the Babbington pits were surface workers. The company appears to have brought pressure to bear on the offenders with the result that on September 25, 1917 Council resolved that:

> 'Agents report to the effect that the Babbington workmen were now satisfied with the extent to which the Surface workers had joined the Association, and appreciated the assistance rendered by the Employers in this matter, be accepted as satisfactory.'

The Brierley Hill dispute occasioned rather more trouble however. The matter was before Council time after time. References were made to the owners, and to the Coal Controller, but eventually the union's patience was exhausted and an official strike was called in November 1917. The owners then promised to refuse employment to non-unionists and on this undertaking work was resumed. The cost of the strike to the union was about £4,900.

C

APPENDIX 'A'

AVERAGE EARNINGS PER SHIFT OF WORKMEN IN (A) THE MIDLAND COUNTIES DISTRICT (i.e. ROUGHLY THE EREWASH VALLEY AND DERBYSHIRE) AND (B) THE NOTTINGHAMSHIRE AND EREWASH VALLEY DISTRICT (MAINLY THE LEEN VALLEY PITS)

Class of Workmen Underground (Adults)	District (A) June 1914		District (A) Nov. 1918		District (B) June 1914		District (B) Nov. 1918	
	s	d	s	d	s	d	s	d
1 Piece work coal getters	9	10½	17	7¼	9	8¼	17	8¼
1 Coal getters on day wage	6	10¼	13	8¾	8	2¼	14	1¾
3 Putters, fillers, hauliers & trammers	6	9¼	12	8½	5	8½	11	2
4 Timbermen, stonemen, brushers & rippers	7	0¼	13	3	7	2¾	13	6¾
5 Deputies, firemen and examiners	8	0	15	5	8	2½	15	3½
6 Other Ugd. labour	5	8	10	10	6	2¼	11	3
Total Adult Labour:	7	9	14	7¼	7	11¾	14	6¾
Surface (Adults)								
7 Winding Enginemen	8	2½	14	5½	8	6½	14	7¾
8 Other enginemen	5	7¼	11	1½	9	6¼	12	9¾
9 Stokers & boilermen	5	1½	10	10	5	9¼	11	1¾
10 Pitheadmen	5	4¼	10	5	6	0	11	8¼
11 Screenhands	4	10¼	9	11	5	5¾	10	10½
12 Tradesmen (skilled)	5	11	11	1½	5	6¼	11	3¼
13 Other Surface labour	4	6	8	4¾	5	0¾	10	7¾
Total Adult Labour:	5	2	10	0	5	7¾	11	1½
Grand Total Adult Labour:	7	2¼	13	5¾	7	2¾	13	6
14 Youths & Boys (Underground)	3	6¾	7	4	3	11	7	9½
15 Youths & Boys (Surface)	2	5¾	5	3¼	2	6¾	5	7½

Source: Coal Industry Commission Reports, Vol. III, pp. 99 & 107. The Ministry of Labour's Cost of Living Index Figure (July 1914 = 100) stood at 200 in June 1918 and 220 in January 1919.

APPENDIX 'B'

MINERS' FEDERATION OF GREAT BRITAIN

CONCILIATION BOARD, FEDERATED AREA

March 12, 1915

Proposals for the Renewal of the Conciliation Board
recommended to
the Coalowners and Workmen in the Federated Area

1. The establishment of a new basis. The new basis to be the price or rate being paid at each and every colliery, for each and every class of work, when 50 per cent on the 1888 basis was the rate of wages in operation. This was in December, 1911. Where a price or rate has been fixed since December, 1911, such price or rate shall be adjusted in accordance with this and the following clause.

2. Existing price lists shall not be interfered with in any way, except that where the rate on the price list is the rate of 1888, or the rate of 1888 plus percentages or additions added since 1888, the rate of December, 1911 shall be substituted for the rate shown on the price list, and the new basis.

3. The new Board shall commence on the 1st May next, and continue until the 30th April, 1918, and thereafter until determined by either party by a three months' notice.

4. The minimum to be 10 per cent on the new basis, as provided for in clause 1, and the maximum to be $23\frac{1}{3}$ per cent on the new basis.

5. The present Rules of Procedure of the Board, subject to the alterations as now agreed to by both parties, shall apply to the new Board.

6. That the resolutions of the 21st October, 1912, the 6th January, 1913, and the 15th April, 1913, applying the advances given on those dates to the minimum wage rates then existing, to remain unaltered, and apply to the minimum wage rates now existing and, subject to the following proviso, shall apply, as well as any future adjustments of wage rates by the Conciliation Board to any minimum wage rates existing at the time of such adjustment. Provided that, in the event of an application to any Joint District Board for an increase or a decrease in existing minimum wage rates, in the comparison between the then minimum wage rates and the then average rate of wages, the average rate shall be taken to be the amount then prevailing less any

percentage additions made therefore since the existing minimum wages rates were fixed; it being the desire and intention of both parties that in fixing future minimum wage rates, the advances given by the Conciliation Board on and since October, 1912, and those which may be hereafter given, shall not be duplicated.

THOMAS ASHTON,
Miners' Secretary.

APPENDIX 'C'

WAGES AND PROFITS 1913-1918

TABLE—SHOWING WAGES AND PROFITS
(NATIONAL FIGURES)

Year	Average Pithead Price	Wages per Ton	Royalties per Ton	Profits per Ton	Tonnage Raised	Persons Employed	Yearly Avg. Wage	Yearly Profit & Royalties
	s d	s d	s d	s d			£	£
Avg. 1909–								(1913 only)
1913	8 8¾	5 5¾	0 5½	0 11½	269,589,210	1,058,140	82	28·0 m.
1914	9 11¾	6 2¾	0 5½	1 2	265,664,393	1,037,700	79	21·5 m.
1915	12 5½	7 9½	0 5¾	1 8¼	253,206,081	935,300	105	27·4 m.
1916	15 7¼	9 9	0 6	2 11¾	256,375,366	980,600	127	43·8 m.
1917	16 8¾	10 5½	0 5¾	2 2¾	248,499,240	1,002,100	129	33·7 m.
1918	20 4¼	13 3½	0 6¼	2 2½	227,748,654	990,300	159	35·5 m.

Sources: Colliery Year Book 1951, p. 537; Coal Industry Comm. Reports, Vol. III, p. 7 (1919).

'Real' earnings, that is, earnings measured in terms of the goods money will buy, can be calculated in a rough and ready fashion by making use of the so-called Cost-of-living Index. Taking July 1914 as 100, the index figure rose as follows:

EARNINGS INDEX AND COST-OF-LIVING INDEX

Date	Total Average Earnings per Man	Earnings Index	Cost-of-living Figure (Avg. for Year)
1914	£79	100	100
1915	£105	133	117
1916	£127	161	140
1917	£129	163	183
1918	£158	201	192

Source: Ministry of Labour Gazette.

APPENDIX 'D'

CHANGES IN PERCENTAGE ADDITIONS TO BASIS WAGES IN FEDERATED DISTRICT 1893–1913

When the Conciliation Board for the Federated District came into existence, in 1893, the miners were being paid the 1888 basis rate, plus 40 per cent. Thereafter, the following changes took place:

PERCENTAGE VARIATIONS ON '1888' BASIS WAGE

Date	Percentage Advance	Percentage Reduction	Addition to 1888 Basis
1894 August	—	10	30
1898 October	2½	—	32½
1899 April	5	—	37½
1899 October	2½	—	40
1900 January	5	—	45
1900 October	5	—	50
1901 January	5	—	55
1901 February	5	—	60
1902 July	—	10	50
1903 December	—	5	45
1904 August	—	5	40
1907 January	5	—	45
1907 May	5	—	50
1907 September	5	—	55
1908 January	5	—	60
1908 September	—	5	55
1909 March	—	5	50
1912 October	5	—	55
1913	5	—	60
1913	5	—	65

THE UNEASY PEACE

1. *The Union Swings to the Left*

The Great War of 1914 to 1918, like the more recent conflagration, generated a strong leftward swing of the political pendulum. The beginning of a war produces a wave of patriotic fervour which can be characterized as a swing to the right. This was true of the Boer War and of the Second World War, but was especially marked in 1914. Keir Hardie and the anti-war party found themselves isolated, their troops having responded to the call of a recruiting sergeant whose trumpet-call was louder than their own. However, as the war progressed, a sense of frustration and disillusionment grew: a feeling that 'this sort of thing mustn't be allowed to happen again' made itself evident. The Labour Party, which had been divided during the early years of the war, formulated a comprehensive programme for the post-war period and its organizational structure was overhauled. In 1918, the Party committed itself to socialism for the first time, and in so doing severed the Liberal connection to which it owed its earlier successes.

The general left-wing tendency made itself felt inside the Nottinghamshire Miners' Association. Twenty years before, J. G. Hancock, the Gladstonian Liberal, had epitomized the union's character and outlook; now he found himself becoming more and more isolated. On the other hand, extremist opinions found widespread support and the union's Council took on a syndicalist tinge. This was largely the result of propaganda work carried out by a small group of people in the Mansfield area, under the leadership of an Irish exile named Jack Lavin. Lavin was educated at an Irish University and was intended for the priesthood. However, he lost his faith and decided to go to sea. After a time at sea, he settled in San Francisco and was there during the earthquake and fire of 1906, when most of his personal possessions were destroyed. In America, Lavin came under the influence of Eugene Debs and Daniel de Leon, and he became an active member of the Socialist Labour Party of America: a body which advocated revolutionary socialism as opposed to social reform. He was associated also with the Industrial Workers of the World, a trade union organization established in Chicago in 1905 under SLP influence. The IWW (popularly known as the 'Wobbly Willies') were opposed to the Craft Union structure of the American Federation of

Labour and insisted that workers should be organized instead by industries.

Lavin came to England shortly before the first World War and took employment in the Yorkshire coalfield, where he became notorious for his advocacy of militant industrial unionism. In accordance with his principles, he refused to accept trade union office and he also refused to stand for Parliament, although he was strongly urged to let his name go forward for the Pontefract Constituency.

In 1915 or 1916, Lavin left Yorkshire and took employment at Welbeck Colliery. Here, he speedily built up a small but influential group of supporters who formed a branch of the Socialist Labour Party of Great Britain. Meetings were organized by this group on Mansfield Market place and at various centres in North Nottingham-shire. Their chief thesis, as Herbert Booth says in a letter to the author, was 'That all trade union leaders are corrupt and traitors to the working class.' Lavin, like many of his countrymen, had great personal charm and he enjoyed considerable popularity in the district. His views gained wide acceptance and his influence lived on after his untimely death from tuberculosis in August 1919. The SLP Group founded by Lavin and Owen Ford became a branch of the Com-munist Party of Great Britain on its formation in 1920.

Another left-wing group had as its leader, Herbert Booth, who had been the Association's first student at the Central Labour College. Booth left the Labour College in August 1914 and was concerned to find that a move was afoot to take the Association out of the MFGB Political Fund. As we saw in our last chapter, together with W. Askew, a fellow member of the ILP, Booth fought strenuously to frustrate this move. Meetings were held all over the County, with Askew as Chairman and Booth as speaker. This campaign had Spencer and Hancock worried—so much so that the former wrote a newspaper article directly criticizing Booth and his associates. Before long, Booth had assisting him a committee of keen socialists who had attended classes at which he was the tutor. Among them were Askew, Harry Alcock of Rufford, George Raynor of Pinxton, Tom Mosley of Gedling and Jack Smith of Hucknall who later became Agent to the Leicestershire miners. This committee decided to widen the campaign by bringing in other issues: abolition of the 'butty' system; pithead ballots for the election of branch officials; and elections at three-year intervals for full-time officials. In addition to their Sunday meetings (which brought them into bad odour with the strict nonconformists) they issued 30,000 copies of a leaflet which asked the miners to demand a ballot vote on the question of whether to continue with political affiliation to the MFGB. As we saw in our previous chapter, this particular issue was brought to a successful conclusion, but the others proved much harder nuts to crack. Later, the Committee added another point to its programme: the abolition of forks or screens which were used at many pits instead of shovels to ensure that only lump coal was sent out of the pit.

This left-wing Committee had a great deal of popular support, but it generated some powerful opposition from the Liberal element, which was still strong in many areas; the prosperous 'butties'; the Socialist Labour Party group led by Jack Lavin, and the leading officials of the NMA. Of these, Spencer was much the most formidable. Spencer attacked the group in Council, in the Press, and on public platforms, until eventually, in 1917, he and Booth debated the issues between them at the Eastwood 'Empire' before an audience of 1,700 miners. At the end of the debate a vote was taken. Spencer received eleven votes, the remainder of the audience voting for Booth's practical left-wing programme.

One point in this programme—the election of branch officials by ballot—was successfully pressed home by 1920. By this time, however, Herbert Booth was no longer in Nottinghamshire. In 1918, he was elected full-time Agent of the Forest of Dean miners, but he returned to the County in April 1922 when he was elected checkweigher at Annesley Colliery. However, Booth's 'ginger' group continued to make itself felt in the immediate post-war period; when the Liberals and Lib-Labs. were, for a time, completely swamped.

The following resolution, adopted by Council on January 28, 1919, is indicative of the change in the Association's political complexion:

'That we petition the Government to discharge immediately all men transferred to classes 'W' and 'Z'; to withdraw all British troops from Russia, to repeal at once the DORA,[1] to take in hand the immediate Nationalization of all Coal Mines and that we strongly protest against the handing back to private ownership of Governmental Ships and Shipyards.'

The views of the Association on demobilization were conveyed to the MFGB Special Conference held at Southport on January 14, 1919, by George Spencer, who said:

'... this Federation might very wisely raise a protest against any men being retained at the present time in the Army for any other purpose than that of the military exigencies depending upon the war we set out for in the first instance, and no man ought to be kept in the Army or retained in the Army for the purpose of imposing the will of a country on another country in relation to its own internal affairs.'

He went on to say that, whilst he did not think it to be 'to the best advantage of either Russia or Germany evolving its form of Government by force rather than by intelligence' this did not give us the right to interfere with their internal affairs.

At a further Special Conference held on Wednesday, April 16th, Spencer pointed out that we had 264,000 British soldiers still under arms in Germany, whilst it was officially admitted that 'the German is incapable of any further effort for the present'. The responsibility for any 'rioting, Bolshevism, or anarchy, or disorder in Germany must rest with the (British) Government that refuses to lift the blockade and starves the people.' Spencer also called for the withdrawal of the 44,000-strong British garrison in Ireland which, he said, was there for 'the purpose of preventing the very principle for which we have been fighting—self-determination'. Again, at the Annual Conference held at Keswick, Spencer declared that Nottinghamshire was 'in favour of taking action . . . in reference to the blockade, conscription and to Russia in particular'. Later in the debate, Spencer made it clear that, although there was some division of opinion among the Nottinghamshire delegation, he was in favour of resorting to 'direct action' in support of the union's political programme.

The demand for direct action was submitted to a conference of the Triple Alliance, where it met with the opposition of J. R. Clynes who thought that a strike undertaken for political purposes would inevitably lead to 'a state of starvation and certainly a state of riot'. The Conference decided to refer the question to the membership of the three organizations which composed the Alliance.

In Nottinghamshire the leadership was divided on the 'Direct Action' issue, and indeed, Mr W. Carter, MP, spoke in opposition to the proposal at Branch meetings. For this he was mildly censured at the MFGB Conference held on September 3, 1919. However, at this Conference, Mr Spencer, quick to respond to the shift in the political outlook of the membership, also came out in opposition to the proposal to take a strike ballot on political issues. He said that Nottinghamshire had already taken a ballot vote on this question and that some lodges had been much less militant than they were expected to be. The reason for this was not far to seek: the miners were now faced with issues of more direct importance to their industry, and in particular the demand for nationalization of the mines, with which we shall deal later. Mr Spencer said that, whilst he saw nothing wrong with direct action, the Federation should save this weapon for use in connection with the nationalization campaign, rather than dissipate its resources over a less important question.

On July 28th, Council resolved that '. . . we withdraw all War Loans as a protest against Allied intervention in Russia', but the Trustees opposed this step because of the capital loss which would result from selling the Bonds instead of holding them to maturity. The trustees refused to carry out their instructions, and indeed, decided to re-invest the £20,000 Exchequer Bonds in a further issue of Government Stock upon redemption.

It would appear from the Minutes that Council was, at this time, more militant than the bulk of the membership; since, despite the lukewarm response to the ballot on direct action in connection with the War of Intervention in Russia, Council continued to pass strong resolutions on the matter. Thus, on December 29, 1919, it decided to support the 'Fight the Famine Council', whose policy it was to secure the 'raising of the Russian Blockade and a modification of certain economic clauses in the Versailles treaty'. On May 31, 1920, Council instructed its representatives to the MFGB Annual Conference 'to move that the Conference call for a special Trade Union Congress for the purpose of considering the question of a National Strike to force the British Government to take action to put an end to these new wars, and that we congratulate the dockers in refusing to load or unload ships with munitions'. At its next meeting, Council went on to mandate its delegates to the Trade Union Congress to 'vote in favour of using the full industrial weapon to end wars in Poland, Ireland, and war in any other part of Eastern Europe'.

Of course, the Union's attitude to demobilization was not solely due to its concern over intervention in Russia's internal affairs. It was concerned with the possibility of unemployment resulting from the expected influx of ex-servicemen; and various means of avoiding this were canvassed. The demand for a six-hour day was partly based upon the need to find work for miners returning from the forces; and the local campaign for the abolition of piece-work and the substitution of all day work was seen in the same light.[2] 'Now, the question of nationalization and hours,' said George Spencer at Southport on Wednesday, January 15, 1919, 'so far as this discussion is concerned, are means to an end. That end, I take it, this morning, is the finding of suitable employment for men who are being demobilized. . . . Now following on the shortening of hours, in my opinion, must be the question of the day's wage system for all men working in and about mines.' Spencer appeared to be arguing that some of the demobilized men might not be able to earn a fair day's pay on contract work, and for this reason, the owners might be opposed to the day-work system so that it would ultimately be necessary to nationalize the mines in order to ensure employment for the partly disabled.

The demands for mines nationalization and a six-hour working day were, of course, part of the post-war programme of the Miners' Federation of Great Britain and local events must be seen against the national background. We spoke earlier of the change in the climate of opinion during the war. This change was undoubtedly due, in part, to the psychological effect of the wartime (and immediate post-war) shortage of labour. As in the more recent war, the men had the feeling that they held the whip hand. Of the traditional incentives, the carrot was bigger and juicier than ever before, whilst the stick was temporarily out of sight. Unfortunately, the miners' leaders of the early 1920s tended to allow their thinking to be coloured by the circumstances of 1914–19. In consequence, their ideas tended to become more and more divorced from reality; and the division of opinion culminating in the Spencer split arose partly from the fact that some people, of whom Spencer was one, learned earlier than the rest that conditions had radically changed and that what appeared practicable in 1919 was so no longer. In 1919 nationalization appeared to be almost inevitable. The men held, as we have said, the whip hand; the industry was shown to be badly organized; and the distributive side of the trade was shown to be in a state of chaos. Further, the public was appalled by the contrast revealed during the sessions of the Sankey Commission between the wretched living conditions of many mining families on the one hand and the exorbitant wartime profits and royalties earned by the owners and landlords on the other. However, by employing a policy of studied procrastination, the Government was able to stifle the demand for nationalization until the psychological moment was passed. Then, when the battle was already lost, the Trade Union Movement opened its 'Mines for the Nation' campaign. But perhaps we ought now to retrace our steps to the beginning of 1919, in order to make the sequence of events clear.

2. The Nationalization Issue

In January 1919, the MFGB formulated a series of demands: an increase of 30 per cent on Gross Wages (excluding War Wage); a six-hour day; full maintenance at Trade Union rates of wages for mineworkers unemployed through demobilization; and nationalization of the mining industry.

On February 10th, the Government offered to increase the War Wage from 3s 0d to 4s 0d a day, and to set up a Committee of Inquiry to consider the question of nationalization and the position of the coal trade generally. This offer was rejected by the union. Instead, a national strike ballot was taken and this resulted in a majority of over half-a-million in favour of a stoppage over the issues mentioned in our previous paragraph.

On February 24th, the Coal Industry Commission Bill, which sought to establish a Royal Commission to enquire into such questions as miners' pay, hours of work, working and living conditions and, most important of all, the organization of the industry, was introduced into Parliament. The Conference of the MFGB, meeting in London, on February 26th and the succeeding days, decided to participate in the work of the Commission provided that the union could be given adequate representation on it. Lloyd George had promised that the Royal Commission would be required to present an interim report by March 20th and, because of this, the Federation agreed to suspend strike notices until March 22nd.

The Coal Industry Commission held its first meeting on March 3rd. It was composed of the following gentlemen:

Hon Mr Justice Sankey (Chairman);
Mr Robert Smillie;
Mr Herbert Smith;
Mr Frank Hodges; and
Sir Leo Chiozza Money (nominated by the Miners' Federation of Great Britain)
Mr R. H. Tawney; and
Mr Sidney Webb
(being Government nominees agreed to by the MFGB);
Mr Arthur Balfour;
Sir Arthur Duckham; and
Sir Thomas Royden
(being Government nominees);
Mr Evan Williams;
Mr R. W. Cooper; and
Mr J. T. Forgie
(representing Coal Owners).

The Commission took evidence from miners, miners' wives, coal owners, royalty owners, mining engineers and experts of one kind or another, and on March 20th issued not one, but three interim reports. One, signed by the MFGB side of the Commission, advocated an increase in gross wages (excluding war wage) of 30 per cent; a shortening of the working day from eight hours to six hours and nationalization of the industry.

The 'Sankey' Report, signed by the Chairman of the Commission and the three Government nominees, recommended a reduction in the length of the working day, of one hour with effect from July 16, 1919, and a further reduction of one hour, depending upon the economic position of the industry, with effect from July 13, 1921. It also recommended a forty-six-and-a-half-hour week (exclusive of mealtimes) for Surface workers, an increase in wages of 2s 0d a day (1s 0d a day in the case of people below sixteen years of age) and a radical alteration ('either nationalization or a method of unification by national purchase and/or by joint control') in the organization of the industry.

The other report, signed by the coalowners' representatives, merely recommended an increase in wages (1s 6d a day for persons of sixteen years of age and over and 9d a day for those under sixteen years of age); a reduction in the length of the working day for Underground workers of one hour; and an eight-hour day for Surface workers.

Mr Bonar Law, speaking for the Government in the absence of the Prime Minister, said that the Government accepted the Sankey Report 'in spirit and in letter'. He confirmed this in a letter to Mr Hodges, General Secretary of the MFGB, dated March 21, 1920, which reads:

'Dear Sir,

'Speaking in the House of Commons last night I made a statement with regard to the Government policy in connection with the report of the Coal Industry Commission. I have pleasure in confirming as I understand you wish me to do, my statement that the Government are prepared to carry out in the spirit and in the letter the recommendations of Mr Justice Sankey's report.

'Yours faithfully,
A. Bonar Law.'

This was taken by the miners to be a promise of nationalization. Because of this, the MFGB Conference held on March 21st instructed the Executive to continue negotiations with the Government and, in the meantime, agreed that the men should continue at work on day-to-day contracts. The Government refused to improve on the recommendations of the Sankey Report as to wages and hours and consequently the terms upon which the men eventually voted were as follows:

'The Government, as the result of the Coal Industry Commission, having offered:

HOURS

1. A reduction of one hour per day in the hours of Underground workers from July 16, 1919, and 'subject to the economic condition of the industry at the end of 1920' a further reduction of one hour from July 13, 1921.

2. Forty-six-and-a-half working hours per week, exclusive of mealtimes, from July 16, 1919.

WAGES

3. An increase of 2s 0d per day worked to adult colliery workers and 1s 0d per day worked for colliery workers under sixteen years

of age employed in coal-mines or at the pit-heads of coal-mines. (The above to apply as and from January 9, 1919.)

NATIONALIZATION

In view of the statement in the report of the Chairman of the Commission that "the present system of ownership stands condemned" and that "the colliery worker shall in future have an effective voice in direction of the mines" the Government have decided that the Commission must report on the question of nationalization of the mining industry on May 20, 1919.'

The voting showed an enormous majority in favour of acceptance (693,084 for; 76,992 against). In Nottinghamshire, 30,385 men voted in favour of acceptance and only 1,764 voted against. This result was announced at a Special Conference held at Southport on Wednesday, April 16, 1919, when it was decided that the strike notices should be withdrawn.

The Sankey Commission now entered on its second stage, which lasted from April 24th to June 23rd. At the end of that time, four reports were issued. One was submitted by the Chairman (Mr Justice Sankey); a second was presented by the Labour members of the Commission (Sir Leo Chiozza Money, Robert Smillie, Herbert Smith, Frank Hodges, R. H. Tawney and Sidney Webb). A third report was drawn up by the coalowners and two of the independent members whilst the fourth was signed only by Sir Arthur Duckham.

The Chairman and the Labour members advocated the nationalization of the industry; the other members of the Commission opposed it. However, all four reports accepted the desirability of nationalizing the coal seams and abolishing royalties, and of placing coal distribution in the hands of public bodies.

On July 9th the Government increased the retail price of coal by 6s 0d a ton. In view of the very large margin of profit then being earned, this increase was totally unnecessary and the MFGB took the view that the increase was a deliberate attempt to turn the general public against the miner by saying that the increased price was necessitated by the 2s 0d a day increase in wages and the seven-hour day awarded as a result of the Sankey Report. Then, on August 18th, the Prime Minister made it clear that the Government did not intend to nationalize the mines. Instead, he proposed the establishment of a number of large trusts.

The leaders of the MFGB were shocked to learn that the Prime Minister was now, as they believed, breaking a solemn promise. But George Spencer had indicated his doubts about the Government's intentions as far back as Friday, March 21, 1919 when he said, speaking at a Special Conference of the Federation:

'I want to refer to Judge Sankey's Report [i.e. the interim report] in relation to nationalization—nationalization or a method of unification by national purchase or by joint control. National-ization is the most important question, and I want to know whether the Cabinet or Parliament is prepared to accept the ultimate report from that Commission; if that report is in favour of national-ization; because that is not in Judge Sankey's Report. I take it it means that he is not going to pledge this Commission to national-ization.'

In his reply Mr Smillie, the President of the MFGB and a member of the Sankey Commission, said: '. . . I do not think any Government could stand for a moment if it sets up a Commission and asks for a report and then refuses to accept that report, but that would be a matter entirely for Parliament after hearing the Commission's report. We should have, however, the weapon of the Triple Alliance.'

No doubt, in the heady atmosphere of 1919, Mr Smillie's attitude was understandable. Looked at over a distance of forty years, however, it seems incredible that any trade union leader could expect a predominantly Conservative Government to accept such a report coming from such a Commission. Obviously, a Commission, one half of whose members were virtually nominees of the MFGB, could be expected to produce a strong case for nationalization. Would any right-wing Government deliberately weight such a Commission so heavily in favour of the men's side if it had any intention of giving the Commission a blank cheque? To carry the argument a stage further, the six left-wing and six right-wing members virtually cancelled each other out on the question of nationalization, thus giving the Chairman the casting vote, so to speak. Could any Government be expected to place such a vital decision in the hands of one man no matter how wise or independent he might be? Mr Varley's attitude when the Commission was first mooted was one of incredulity. Speaking at an MFGB Special Conference on Wednesday, February 26, 1919 he said, '. . . what is public opinion going to say to us when they find we would not previously accept the findings of a Committee but we are now prepared to accept the findings of a Commission if we can pack the jury? But there is no guarantee that we are going to pack the jury. The Prime Minister says that there are three different sets of interests involved. There are the miners, the mineowners, and the country as represented by the Government. The miners want half, but we shall not get half. If we do get half, there is a new situation arisen, and that is this, that by the attitude of the Parliamentary party in the House of Commons last night they have placed us in the position of saying we will not accept the findings of the Commission by March 31st, but we will by March 20th, and even on March 20th, the findings of the Commission may be against us.'

By promising a Royal Commission which met all three requirements of the MFGB (namely, that it should consider the question of nationalization as well as wages and conditions; that its composition should be heavily weighted in the union's favour and that it should report quickly) Lloyd George averted a strike at a time when public opinion was favourable to the miners. By conceding the increase in wages and reduction in hours recommended by Mr Justice Sankey promptly, Lloyd George took the edge off the agitation for nationalization in the coalfields; and by increasing the price of coal immediately thereafter, he turned public opinion against the miner. By delaying the day of decision on the nationalization issue to August 18th, the Prime Minister made the possibility of a successful strike to enforce nationalization extremely unlikely. Mr Robert Smillie might, in March, regard the Triple Alliance as a 'weapon'; Lloyd George correctly regarded it, in August, as a weak reed.

The NMA Council, at its meeting on August 25th, resolved 'That this Council strongly protest against the Government declining to Nationalize the Coal Mines as per the Sankey Report'. Three days later, Council adopted a further resolution: 'That our delegates to the London Conference on Wednesday, September 3, 1919, support the Federation in any action they may deem desirable to enforce the Nationalization of Coal Mines as per Lodge vote just given: For 659, Against 56.'

In practice, the MFGB Conference of September 3rd decided not to take industrial action; but decided instead to '. . . invite the Trades Union Congress to declare that the fullest and most effective action be taken to secure that the Government shall adopt the majority report of the Commission as to the future governance of the industry'.

On the day following the MFGB Conference, the Triple Alliance, meeting to discuss an indefinite postponement of the strike ballot to determine whether the Government should be forced by industrial action to withdraw its troops from Russia, demonstrated its uselessness. The delegates indulged in senseless squabbling before agreeing by 182 votes to 45 to an indefinite postponement of the ballot. Clearly, the weapon upon which Mr Smillie had placed such reliance was not going to be of much use in the campaign for nationalization.

The Parliamentary Committee of the TUC, together with the Executive Committee of the MFGB, met the Prime Minister on Thursday, October 9th. At this meeting, the Right Hon W. Brace, MP, for the MFGB, put his finger on the insuperable obstacle standing in the way of union participation in the management of privately-

owned trusts. He said, 'I do not think you ought to ask us to accept
a suggestion such as your Government has made, or such as Sir
Arthur Duckham proposes, because you are asking an organization
which has been established to defend workmen against the power of
capital to take part in a movement which a body constituted such as
ours is cannot accept'. Having effectively ruled out the idea of jointly
controlled trusts, Brace then went on to warn the Government
'. . . that if they go back to the old system they are up against a
something which will absolutely ruin us. Really, after all, part of this
movement (for nationalization) is psychological—a good part of it.
The war has driven us twenty-five years at least in advance of where
we were in thought in 1914. The young men have thought deeply,
and indeed, they are educated.'

Both Mr Bonar Law and Mr Lloyd George dealt with the accusation that the Government had broken faith with the miners in not
accepting the nationalization proposal. Mr Bonar Law said that his
statement in Parliament accepting the report in the spirit as well as
in the letter was merely an acceptance of the interim report and not
of any that might subsequently be made. He went on to say of the
interim report: 'All there is in that is this; that the existing system
stands condemned and must be replaced in some way or another, as
Mr Justice Sankey said, either by control or by nationalization; and
the proposal made by the Government in the speech of the Prime
Minister did endeavour to deal with it in the second of the alternatives suggested.'

Mr Lloyd George said: 'It is suggested that we have signed a bond
as a Government to carry out all the recommendations of the Royal
Commission under all the headings of enquiry, whatever the majority,
even if there is a majority of one: that the Government without exercising any independent authority or independent judgment of their
own are bound to accept every recommendation, every plan, every
scheme put forward by that Commission were it only by a majority
of one, and that a failure to do so is to dishonour our bond'. Lloyd
George went on to say that he had 'never heard of a Government
that has ever taken up that position'; and he described the idea that
the Government, Parliament and the country, were bound to accept
Mr Justice Sankey's recommendation as 'a doctrine which seems to
me to be of so sweeping a character that I cannot imagine anyone
with a sense of responsibility accepting it'. He went on to criticize
Sir Leo Chiozza Money and R. H. Tawney for going into the witness
box and then as members of the Commission 'to decide upon the
quality and the merit of the evidence they themselves had given'.

It was painfully obvious that the Government were not open to
conviction and the TUC, the Labour Party and the MFGB, therefore,

D

decided to launch a joint campaign for Nationalization of the Mines. At its meeting on November 1st, the Council of the NMA had resolved: 'That any large Lodge, or number of small Lodges combined, desiring to hold Mass Meetings to promote Nationalization be requested to inform District Officials of this desire, and also send name of any person they wish to address it in addition to District Officials, and Officials make the necessary arrangements.' The first speaker asked for under this arrangement was Mr Frank Hodges, then General Secretary of the MFGB. Further meetings were held throughout the county, but they appear to have aroused little interest among the general public.

In the early months of 1920, the MFGB decided to raise once more the question of direct industrial action over the nationalization issue with the TUC. The Nottinghamshire Council instructed its delegates to vote in favour of nationalization at a conference of the MFGB held in London on March 10, 1920. Unfortunately, the MFGB Conference was itself deeply divided on the 'direct action' issue, and when the Special Trades Union Congress met on the day following, it decided by 3,732,000 votes to 1,050,000 not to organize a general strike, but instead, to carry on an intensive propaganda campaign. However, political propaganda, no matter how 'intense', could no longer make a live issue out of nationalization. The match was over bar the shouting, with Lloyd George an easy winner.

3. Local Issues

The progress of the nationalization campaign in the Nottinghamshire coalfield, with its alarms and excursions, its conferences, its public meetings and its strike ballots, presents inevitably a confused picture because of the important local issues which were bound up with it. The principal local issues were: a demand that all screens and forks should be taken out of the pits; a determined campaign to end the butty system once for all; and a new county agreement for Haulage Hands and certain other grades.

Filling by screen instead of by shovel results in a drop in earnings since the small coal which passes the screen is left in the waste. For this reason, and also because to leave any quantity of slack in the waste is to encourage 'gob' fires, miners have always opposed screen filling, although they had often been forced to endure it.

On January 11, 1919, Council decided to take a ballot vote to determine whether members were 'in favour of giving fourteen days' notice to terminate their contracts to enforce shovel filling, also to secure for the Main Road Workers an 8s 3d basis rate for all Haulage Chargemen, and 7s 6d for all other workers, and 8s 3d per shift for all Contractors working in abnormal places'. A majority voted in favour of striking and notices were, therefore, handed in on Wednesday, January 29th at all collieries. On January 27, 1919, Council also endorsed the all-throw-in system for work done on contract, subject to certain amendments which the owners wished to make.

Because of the negotiations which were taking place, Council agreed, on February 10, 1919, to recommend the suspension of notices for one month, the men to continue working on day to day contracts meantime. This month was to be a trial period, with shovels being used instead of screens. The Council's recommendation was accepted by the men, 25,949 votes being recorded in favour to 1,944 against. Council also took this opportunity to instruct all branches to operate the all-throw-in system. Negotiations with the owners on the other outstanding grievances (new price-lists for main road workers and clerks, a claim for payment for setting benk-bars, and a make-up rate for contractors of 8s 3d a day) continued, but without result. Consequently, the Association officials asked the Coal Controller to intercede.

At its meeting on March 10th, Council agreed to suspend notices for a further two weeks, in order to avoid clashing with a decision of the National Conference. It will be remembered that negotiations were taking place at this time over the wider issues (e.g. nationalization, a national increase in pay and a reduction in the length of the working day) at national level; and that the MFGB Conference had agreed to suspend the national strike notices until March 22nd. However, on March 18th, Council rescinded its motion to suspend notices and instead instructed delegates to 'inform their respective members to make their working places in the stalls as safe as it is possible, and cease work tomorrow, March 19, 1919'.

On March 19th, the only people at work underground in Nottinghamshire were a handful of deputies, safety men and horsekeepers, who were working by arrangement with local strike committees. However, the strike did not last for long. Negotiations with the owners were speeded up, and on March 27th, the Strike Committee recommended a resumption of work. This recommendation was rejected by 422 votes against 305, but four days later, a return to work was decided upon. The owners had now agreed that shovels should be used instead of forks and had offered a new Wages Agreement for main road workers, and make-up rates of 8s 3d for chargemen and 7s 10½d for others. It had also been agreed between the two sides that all other outstanding issues should be 'placed before the Coal Controller'.[3]

Most of the pits resumed work on Wednesday, April 2nd, and by the following morning they were all back at work.

This did not settle the issue, however, since the Agreement was not formally ratified and in consequence was not fully honoured. The Barber Walker men in particular were up in arms at the failure of their Company to pay the new rates. At Stanton Hill, also, the owners were not paying the revised rates for chargemen, rippers, timberers, and similar categories; and the men there were given leave to tender strike notices by Council.

On August 19, 1919, the permanent officials interviewed Sir Robert Horne, the Minister of Labour, in order to secure his assistance towards finalizing the Main Road and Surface workers agreement. This had no immediate effect, however, and the question was referred once more to the Union's strike committee on August 25, 1919. At the same time, the owners were requested to complete the Agreement, but since they had not replied to this request by November 1st, it was then agreed that a further strike ballot should be taken throughout the County.

The union continued to seek ways and means of securing a peaceful settlement of the dispute and, in November 1919, Mr Spencer once more sought the assistance of the Minister of Labour and the strike ballot was postponed. The negotiations dragged slowly on until in March 1920, the strike ballot was, at last, held. This showed a large majority (14,124 to 3,317) in favour of taking strike action over this issue. Even now, Council decided to defer action for a further month in order to give the Coal Controller a chance to arrange arbitration, with the result that the Minister of Labour appointed Mr W. H. Stoker as arbitrator under the Industrial Court Act, 1919.

The case was argued before Mr Stoker by representatives of the two sides on May 11th, 12th and 27th and on June 8, 1920, and in addition, the arbitrator visited three pits in order to view the working conditions of some of the people involved in the claim. The claim was for increased wages for haulage and other main road workers, underground, surface workers and bank workers.

During the sitting of the Court, the Owners' Associations and the union agreed on the rates to be paid to underground haulage workers. For adults, the shift rate was to be 6s 8d basis, plus current percentage.

The claim in respect of shop men and tradesmen was for an increase of 6d per day on the basis rate; and, in respect of boys working in the shops, for pay on the same scale as for boys employed on the bank. The union spokesmen pointed out that certain grades of enginemen represented by the Enginemen and Firemen's Union had received an increase of 6d a day in their basis rates, by an agreement reached on March 21, 1919, and the present claim was based partly on that, partly on a recent advance to similar workers in the South Yorkshire coalfield, and partly 'on the general financial position of the coalfields in Nottingham as compared with coalfields in the other parts of the Kingdom as evidenced by the statistics appended to the Report of the Sankey Commission'.[4] The rates paid to skilled shopmen in South Yorkshire were shown to be 6d a day higher than those paid in Nottinghamshire, whilst other surface workers in South Yorkshire were paid 2d or more a day above their counterparts in this county.

The claim for banksmen was for an additional 6d a day on the basis rates for all aged twenty-two and over in order to re-establish the differentials disturbed by the enginemen and firemen's agreement of March 21, 1919.

On June 16th Mr Stoker issued his Award, which substantially met the Union's claim. The new rates for shopmen ranged from 6s 2d for fully skilled men (e.g. fitters, blacksmiths, joiners and bricklayers) to 5s 6d for lower grades of craftsmen (e.g. tram repairers and lamp repairers). The rate for semi-skilled men (e.g. blacksmiths' strikers and fitters' assistants) was 5s 4d, and the rate for general labourers was 5s 2d.

Further, the arbitrator raised the adult banksmen's rate to 5s 6d, and he agreed that boys employed in shops should be paid the appropriate bank rate according to age.

The Award, which came into effect on June 1, 1920, settled all the surface rates except for brickworkers. The arbitrator was unable to fix rates for these people because of insufficiency of data. One finds this difficult to understand since a great deal of attention had been paid by Frank Varley to the brickworkers' terms and conditions, and in the previous year, there had been a very lengthy strike at the Babbington Brick Works, which was satisfactorily settled in November 1919. It may be, of course, that the union officials were unable to produce evidence as to the profitability of colliery-owned brickworks; and they may also have found difficulty in obtaining details of earnings in brickyards which were not associated with collieries. Alternatively, they may have withheld any data they possessed on these two subjects if such data supported the owners' case rather than their own.

Meantime, the debate on methods of payment for stallmen continued. It will be remembered that a majority of the men had voted for the introduction of payment by the day in place of the butty system in 1918; but the men themselves were badly divided over this issue. In the 1918 ballot, 7,499 men had voted for the day wage system, but 3,579 had voted in favour of retaining the butty system; and a further 4,698 had voted for the all-throw-in system. It is safe to assume that most of the butties voted in favour of their own continued existence; and they were more powerful than the day-men.

For one thing, the checkweighman was their employee; and for another, they were usually on good terms with the colliery officials. Indeed, it is notorious that, at many pits, butties 'kept the gaffers sweet' by tipping them heavily. Further, the owners preferred the butty system because it resulted in a high output for a relatively low labour cost; and finally, inertia favoured the existing method.

The ballot vote of 1918 did not, therefore, result in the introduction of a day wage system. Instead, the union and owners compromised with the all-throw-in system under which all the adults employed on a face (or in a stall) share equally in the contract earnings. At the time of which we are now writing, this was the official system; but many pits refused to work it. There were some collieries where those in favour of the day wage system were sufficiently powerful to enforce their views for a time; and there were other pits where the butties, with the active collaboration of their employers, were able to maintain their privileged position.

We saw earlier in this chapter that Council instructed all Lodges to operate 'all-throw-in' on February 22, 1919. However, on July 28th, we find Lodge branch asking for permission to draft a modified butty system for their own pit, but permission was refused. There is evidence, however, that Council's instructions were being ignored at some places since, on January 26th the secretary was instructed to write to the owners 'for a meeting to be held for the purpose of coming to an agreement upon the all-throw-in system'. This motion was moved by the Rufford delegate, and seconded by the Summit delegate, both of whom represented left-wing branches. The two Kirkby branches, Summit and Low Moor, were responsible for a further motion at the next meeting of Council, which reads: 'That steps be taken to abolish all forms of piece-work, and that this resolution be sent to the MFGB', whilst on April 1st, Radford and Cinderhill introduced a motion: 'That a ballot of the county be taken on the direct question as between day work and the all-throw-in system.' At its next meeting, Council agreed to refer this question to the MFGB Conference.

Nottinghamshire had, of course, put this question forward before. At the National Conference held in July 1919, Nottinghamshire had a motion advocating the abolition of piece-work on the agenda but the Executive Committee asked for this to be withdrawn because, as the President (Mr Bob Smillie) said: 'We don't want to do anything at the present time which will interfere seriously with the output.'

The MFGB had plenty to worry about with national issues, and the butty system was regarded as a relatively minor problem affecting only a few areas. Mr Frank Hodges explained to a Conference of the Triple Alliance, held on August 31, 1920, that: '. . . not only has the Federation generally, but every one of our workers has been so keen in years gone by on the abolition of that system that it does not exist as a system in any district, save in one or two isolated cases.' In saying that 'every one of our workers' had been keen on the abolition of the butty system, Hodges was going very much too far. Another delegate from the MFGB said that his brother, a butty, 'recently spent three days at York races, and returned home last Friday in time to pay his men and to pick up over £12 for the week'. This individual may have been grieved at the existence of a system which enabled him to earn £12 a week for doing nothing whilst his day men sweated down the pit, but if so, then no doubt he managed to bear his misfortunes bravely.

The Gedling men struck work in November 1920 over the continued existence of butties. The men, who were out for a week, demanded the introduction of the all-throw-in method of payment but they did not finally get their way until the Digby Colliery Company was absorbed by B.A. Collieries, almost two decades later.

Mr J. G. Hancock, the Union's veteran official, was opposed to the introduction of day work. Equally, he was oppposed to nationalization. The men at Cinderhill wished to abolish the Contract payment system and, in February 1920, they sought the Association's permission to come out on strike. As part of the settlement, the Babbington Company agreed to accept the day work system in the Deep Soft Seam (the other two seams were already on day work). However, they fought a rear-guard action, and in this, they had the support of Mr Hancock.

In a letter dated March 12, 1920, the General Manager of the Babbington Coal Company tells Mr Hancock: 'I am very pleased to hear from you in your letter of the 11th inst., that you are opposed both to Mine Nationalization and also to the Day Work System.' The letter alleges that the average output per coal getter per day had fallen from 2 tons 6 cwt to 1 ton 15 cwt.

At this stage, J. G. Hancock was an anachronism. He was a Nonconformist radical of the type common in the last quarter of the nineteenth century. As we saw in our first volume, he contested Mid-Derbyshire as a Lib.-Lab. candidate in 1909. At this election he had the support of A. B. Markham, the coalowner Liberal MP for Mansfield,[5] thus continuing the policy initiated by William Bailey a quarter of a century earlier. Hancock's secret correspondence with the General Manager of the Babbington Coal Company was all of a piece with his earlier associations. Dr J. E. Williams says of the 1909 election: 'Both Markham and Hancock attempted to obscure any distinction between Socialism and Liberalism which might have existed in the minds of the electors.'

Hancock had been a most unsatisfactory Labour MP. He accepted the Labour Whip but regarded himself as a Liberal. In 1913 the Divisional Labour Party protested against Hancock's attachment to the Liberal Party. There can be no question that, at that stage, Hancock had the support of a majority of the miners of his constituency besides enjoying the support of the NMA. The war years, however, changed all that. Hancock held the seat as a Liberal from 1914 (when the Labour whip was withdrawn) to 1922, when he was defeated in a three-cornered contest.

As we have seen, the political complexion of the NMA changed completely between 1914 and 1919. The post-war world found Hancock exposed as an anti-socialist; and because of this, the right wing of the union tended to crystallize round him. The importance of this will become clear in later chapters.

Naturally enough, Mr Hancock's attitude met with the disapproval of the left wing and attempts were made to remove him from office. A motion: 'That the question as to whether Mr J. G. Hancock should be called upon to tender his resignation to the next Council meeting be referred to the Lodges for their consideration' was sponsored by Mansfield and Newstead in May 1920. At the following meeting of Council, thirteen votes were recorded in favour of demanding Mr Hancock's resignation and twenty-five against.

That was not the end of the matter however. Four months later, Silver Hill and Bestwood proposed that the resignation of Mr Hancock should be the subject of a ballot vote of the whole of the membership; and in November Council agreed by 483 votes to 382 that such a ballot should be held.

The Committee appointed to supervise the ballot reported to Council on January 31, 1921 that they had interviewed Mr Hancock, and in view of his attitude had taken legal advice. The advice given to them was to the effect that the ballot, if taken, 'would not alter the position of Mr Hancock in any degree!' The Committee therefore saw no point in proceeding with a ballot vote the result of which could not be enforced. The whole matter was therefore referred to the Staffing Committee which decided to 'downgrade' Hancock from Secretary to Agent. G. A. Spencer became General Secretary, F. B. Varley, who had been a full-time official since May 26, 1919, became Financial Secretary, whilst C. Bunfield took over the Treasureship from Lewis Spencer, who now entered into semi-retirement as caretaker. J. G. Hancock and W. Carter were Agents without specific offices.

Later on Spencer and Varley were to be regarded as men of the right, but at this stage they were—in appearance at least—moderately left wing. At a Conference of the MFGB on Tuesday, January 14, 1919,

Varley said: '. . . We have implicit faith in the power of our industrial action with regard to the shortening of hours and nationalization'; whilst later at the same Conference George Spencer said, speaking of the demobilization of mineworkers: 'The final solution of this question lies in the State ownership of the mines because the mine owners may not be able to meet this question in a manner which will give justice to those returned, but this cannot be said of the State if it assumed, owned and controlled the mines, because it will not then be a question merely of profit but will be a question of a national obligation.' Again, at the Southport Conference of the MFGB, held on Wednesday and Thursday, February 12 and 13, 1919, Varley thought that the Conference might well decide on strike action (because of the Government's unsatisfactory reply to the Union's demand for higher pay, shorter hours, nationalization, and support for demobilized miners) rather than go to the trouble of taking a ballot vote of the membership. The reasons he advanced for this are revealing. He said: '. . . I was wondering whether we require to face a ballot. We have had some wonderful happenings in Nottingham. We had 20,000 men on strike for four days, arising out of a question which originally was bound up with the demobilization problems, and after having satisfactorily settled that matter they drew up a programme. . . . I am afraid in the County of Nottinghamshire, where our men are already taking alarm, that we shall have the utmost difficulty because we have spasmodic strikes now. We have two pits out today.' Here, we have the reason for the left wing stand of George Spencer and Frank Varley: they reflected the temper of the men as the chameleon reflects the colour of its surroundings. Later, as it became more and more apparent that the Government and the owners were recovering their grip on the situation, they recognized the hopelessness of the extremist position; but in these early years they epitomized the optimism and determination of the men, whilst Hancock clung to the outlook of an outmoded philosophy. In 1919 Hancock provided the right with its rallying point: he was, so to speak, a Spencerite long before Spencer himself became one.

4. *Piece-work Pay Adjustment*

Earlier in this chapter, we noted that the length of the working day was reduced by one hour as a result of the Sankey Report. Of course, this reduction in hours, unaccompanied by mechanization, could be expected to lead to a fall in output per man, and because of this, the Federation asked that men who were paid on contract should have an appropriate increase in their contract rates.

This question was being debated at the same time as the increase of 6s 0d per ton in the retail price of coal. At a meeting of the MFGB

Executive Committee, held on Wednesday, July 16, 1919, resolutions on these two subjects were formulated for the Agenda of the coming National Conference. The first resolution asks that '. . . the maximum figure to be accepted in any given district for the reduction of one hour per working shift shall be 14·3 per cent . . .' The second resolution declares that the proposed increase of 6s 0d per ton in the price of coal is '. . . not necessary and should be avoided' and goes on to argue that the introduction of the economies proposed by the Sankey Commission, together with the re-organization which nationalization would make possible could '. . . make the industry self-supporting, without additional charges to the consumer'.

The proposed advance for piece-workers was discussed at national and local levels. These discussions illustrated all too clearly how very shaky the arithmetic of negotiators can be. Frank Hodges, at a meeting between the Prime Minister and the MFGB Executive Committee, discovered the wonderful truth that 'If you have ten tons of coal at £1 a ton, that gives you £10. If you reduce it by 10 per cent you get nine tons of coal at £1 a ton, that is £9. But if you put 10 per cent on to your £9, you do not get £10, you only get £9 9s 0d.'[6] The Prime Minister and Bonar Law were no better.

After interminable discussions, a formula was finally agreed which gave the Nottinghamshire men an increase of roughly 14·2 per cent in their piece-rates. This did not end the matter however, for on September 25, 1919, the Coal Controller wrote to the NMA querying the increase of 14·17 per cent fixed for Nottinghamshire in accordance with the National formula. Apparently, however, the Nottinghamshire officials were able to satisfy the Controller, since we hear nothing further of the matter. (Perhaps the Controller was also weak on simple arithmetic.)

Of the 6s 0d increase in the price of coal, many harsh words were spoken, but the strike which the militants felt should be organized did not take place.

5. *The Uneasy Peace*

This was, indeed, a period of uneasy peace. As in 1913 and early 1914, feeling ran high over all sorts of issues. Dockers refused to load arms for Russia, whilst Policemen and Prison Officers—of all people—came out on strike and were supported financially by other unions, including the NMA. In the mining industry, Deputies and Clerks, normally peaceable people, became militant.

The deputies had their own union of course, but when their members struck work, as they did at Lodge in January 1919 and at Bentinck in April, the miners also had to stop work.

The clerks were somewhat divided. Many of them—particularly

at the pits near to Nottingham—joined the NMA, whilst at some other pits, another union (what is now known as the Clerical and Administrative Workers Union) recruited members.

On April 7th, Council decided to take action to enforce the pay scale which had been drawn up for clerks and to force clerks into the union. This brought a clash at the Butterley group of pits, where the National Union of Clerks had its stronghold and where it had organized a strike. This body tried to negotiate a working agreement with the MFGB but the Federation Executive turned the proposal down. At its meeting on September 2, 1919, the MFGB Executive had before it a letter from the Right Hon C. W. Bowerman, who wrote on behalf of the Parliamentary Committee of the TUC: '. . . relative to the dispute between the (National Union of Clerks) and the Derbyshire and Nottinghamshire Miners' Associations'. The MFGB agreed to take part in an enquiry into the dispute.

Besides withdrawing labour at the Butterley pits, the National Union of Clerks also decided to call out their members at the Barber Walker pits in July 1919. Council resolved: '. . . That we cannot support such action, and that the officials lend all assistance to the Eastwood Local Officials to prevent any stoppage of work on the part of our men.' Had the National Union of Clerks been successful in stopping production by calling out the office staffs, then some thousands of miners would have been without pay owing to a dispute to which they were not parties. For this reason, the NMA thought it proper to authorize its clerical members at the collieries concerned to continue at work.

A very similar situation arose in October 1919, when the Enginemen and Firemen's Union—which was affiliated to the MFGB—issued strike notices for their members in Nottinghamshire and Derbyshire. Here again, a strike called on behalf of this relatively small section of the men would have resulted in 80,000 miners being thrown out of work.

This question was raised by Mr Varley at a Conference of the MFGB held on Wednesday, October 22nd. In reply, Mr Annable (for the Enginemen) explained that his association had made '. . . an application to the Midland Counties' Coalowners (who covered Derbyshire and the Erewash Valley district of Nottinghamshire) in accordance with Clause 2 of the Sankey Report. The Midland owners agreed for the men to work forty-eight hours, one-and-a-half over the forty-six-and-a-half as set forth, and they would pay a quarter (shift) over—that is, they should have the extra quarter for the forty-eight hours. After that had been agreed upon by the Midland Counties, we sent the agreement to the Nottinghamshire (i.e. the Leen Valley) owners, and they, in turn, by letter on July 29th, accepted that agreement in its entirety and agreed to pay back to July 16th. That went on all right until August 15th when they received a report from the Coal Controller . . . (and) as soon as they got that report, one of the managers said he should not continue to pay, or would rather discontinue to pay, the extra quarter because they had instructions from the Coal Controller that we are not to pay, and eventually the coalowners had a meeting and decided they would cease to pay, and today they are discontinuing to pay the extra quarter for the forty-eight hours, and it is with a view to forcing them to honour the agreement which they signed that the Nottinghamshire enginemen yesterday tendered their notices.'

The Chairman of the Conference, Mr Smillie, urged upon the enginemen the unwisdom of calling out their members without first seeking the co-operation of the two county miners' Unions and of the Federation head office. 'We don't think,' said Mr Smillie, 'there should be a risk of throwing 80,000 men idle until we have had an opportunity of trying to settle.'

This question was referred to the Joint Board of the Nottinghamshire and Derbyshire Miners Associations and the Enginemen and Firemen's Union and a peaceful settlement was eventually arrived at.

The NMA was also in dispute, in a mild sort of way, with the Tramway Workers' Union, who complained that some miners were working on the trams after completing their shift at the pit.[7] Complaints of this sort were often made by the Hairdressers' Federation and the Agricultural Workers' Union, with some justice; but one finds difficulty in accepting the Tramway Union's complaint at face value.

The uneasy peace was also disturbed by the dissensions of Workers' Educational Association supporters on the one hand and National Council of Labour Colleges supporters on the other. The union associated itself with the work of the WEA, but the left-wingers from the Mansfield area preferred their education to have a Marxist flavour. From time to time, the Association sent students to the Central Labour College (the organizing centre of the syndicalists and industrial unionists) for a full-time two-year course. Thus, in 1920, Scholarships were offered to W. F. Paling, of Sutton, G. A. Winson, of Broxtowe and Andrew Clarke, of Rufford.

Of course, the ordinary business of Trade Unionism had to be carried on in the midst of these various dissensions. The union put up its weekly subscription to a shilling a week, so as to support an increase in the allowance to old age pensioner members; a convalescent home, and an extension in the union's political and industrial work.[8] In August 1919, the Trustees considered the question of buying shares in colliery companies, but did not consider the time ripe. Apparently, their views as to the profitability of mining concerns were not so sanguine as those of their public spokesmen.

The nonconformist conscience broke through at the Council Meeting on February 23, 1920, when it was resolved: 'That this Council strongly favours some form of Local Option in the control of the liquor trade.'

More seriously, attempts were made in 1920 to secure an amalgamation of the Nottinghamshire, Derbyshire and Yorkshire Miners' Associations but they came to nothing. One feels that such an amalgamation, had it taken place, might well have changed the course of events. Had the three most productive counties—largely protected by geographical circumstance from the vagaries of the export market—been united during the struggles of the 20s, the extremists of South Wales would not have had their own way so often or so easily. The Miners' Federation of the post-Great War World lacked the ballast which was formerly provided by the old Federated Area. The Nottinghamshire move towards amalgamation of the three counties would have provided such ballast; but the amalgamation did not take place, and when the storms of 1921 and after struck the ship, this lack of ballast made itself felt.

APPENDIX 'A'

BALLOT VOTE ON BUTTY SYSTEM
VOTES RECORDED

Branch	Question 1 Present System	Question 2 Modification Thereof	Question 3 All-throw-in	Question 4 All Day Work
Linby	72	28	178	292
Hucknall No. 2	19	24	78	162
Wollaton	21	20	46	221
Annesley	11	10	140	250
Pye Hill	42	20	36	77
Newstead	120	10	279	389
Cinder Hill	29	26	115	156
Newcastle	15	18	66	75
Huthwaite	202	40	146	532
Bestwood	69	22	120	465
Stapleford	11	9	18	55
Clifton	60	20	36	118
Stanton Hill	39	19	81	157

Branch	Question 1 Present System	Question 2 Modification Thereof	Question 3 All-throw-in	Question 4 All Day Work
New London	61	14	40	49
Brierley Hill	40	10	111	108
Brinsley	10	9	18	59
Bulwell	34	16	58	36
Underwood	33	28	105	93
Hucknall No. 1	22	12	65	227
Pinxton 1 & 2	67	87	144	269
Moor Green	38	27	81	129
High Park	22	11	34	32
Summit	77	67	214	789
Broxtowe	33	21	55	140
Digby	25	3	12	22
Watnall	23	11	76	84
Langton	68	22	83	173
Normanton	27	44	49	62
Bentinck	82	58	148	459
Radford	16	19	25	148
Birchwood	19	20	12	28
Silver Hill	16	19	49	99
Gedling	67	14	242	338
Sherwood	137	64	284	220
Mansfield	138	376	465	76
Pollington	8	2	6	64
Cotes Park	37	38	42	45
New Selston	18	25	28	62
Rufford	73	144	486	51
Welbeck	116	44	294	424
Lowmoor	37	27	100	235
Lodge	11	16	33	29
Total	2,065	1,514	4,698	7,499

APPENDIX 'B'

COPY

THE BABBINGTON COAL CO.,
NOTTINGHAM.

12th March, 1920.

J. G. Hancock, Esq.,
Miners' Offices,
  Old Basford.

Dear Sir,

I am very pleased to hear from you in your letter of the 11th inst. that you are opposed both to Mine Nationalization and also to the Day Work System. I cannot think that the former would be any advantage to the country or even to the miners and experience in the past in that sort of thing has proved that it is very detrimental. The following figures speak for themselves with regard to Day Work as against the Contract System. Taking eight weeks before the Contract System ceased at Cinder Hill the average of coal getters per day in the seam in question was 2 tons 6 cwts., the average for the eight weeks following the day when they started day-work was 1 ton 15 cwts. and it got worse as time went on. You will thus see that there is a drop between the two systems of 11 cwts. Taking a longer period it would have shewn even a greater difference.

I should perhaps point out to you that we have opened out several new seams at these Collieries during the last few years and that at Cinder Hill out of three seams only one, the Deep Soft, was on contract arrangements when the whole Pit decided to go on day-work.

I have not had an opportunity of showing your letter to the Company so perhaps it would be as well for you to keep these figures confidential until I have their consent for them to be made use of in any way you like.

Yours faithfully,
(Signed) E. C. Fowler.

APPENDIX 'C'

THE FUNERAL OF JOHN LAVIN

The following report is taken from the *Mansfield Advertiser* of August 15, 1919.

SOCIALIST SONG AT WARSOP GRAVESIDE

The unusual spectacle of a Red Flag Funeral was seen at Warsop on Saturday, 9th August, 1919, where the interment took place of the late Mr. John Lavin, who worked at Welbeck Colliery, but was widely known as a leading member of the British and International Socialist Labour Party and Workers International Industrial Union.

The coffin, which was draped with the red flag, on which was a wreath of crimson blooms and bunches of roses, was borne from his residence in Alexandra Street by 'comrades' wearing red ribbons in their coats, and they were followed by about 60 men, some of whom came from Manchester, Newcastle, Yorkshire, Warwickshire and Kent.

The journey to the cemetery from the house was nearly a mile and it was necessary for several groups of men to take their turn in carrying the deceased comrade.

There was no religious ceremony at the graveside. First of all Mr. Owen Ford read a long eulogy from a provided form, in which occurred such sentences as the following:

'He fought the good fight of free enquiry, and triumphed over prejudice and the result of misdirected education. He was free from the forces and superstitions of belief. No man was the keeper of his conscience. His religion was of this world, the service of humanity his highest aspiration. No sacred Scripture or ancient Church formed the basis of his faith, and by his example he vindicated the right of thinking and acting upon conscientious conviction.

'For worship of the unknown he substituted duty, and for prayer work, and the record of his life bears testimony to his purity of heart. His end saw him in perfect tranquility, with no misgivings or doubts, no trembling lest he should have missed the right path; he went undaunted into the great land of the departed.'

The coffin was lowered into the grave amid cries of 'Poor Old Jack'.

Mr. Poynton said Lavin was loved the most by those who knew him best, and hated most by those who knew him least. In spite of adversity and the opposition of the forces arrayed against him he stood firm in proclaiming what he believed was in the interests of the class to which he belonged.

The message he left was that all his comrades should work for those interests. He wished to pay tribute of appreciation and gratitude for the work that their friend Lavin had done.

After other comrades had paid their tributes, Mr. Owen Ford read again from the pamphlet, adding 'May the earth lie lightly upon him, may the flowers bloom over his head, and may the winds sigh softly o'er his grave. Peace to him, a long farewell, Jack'.

The coffin, which had brass furniture, bore the inscription:

'John Lavin, died August 6th, 1919, aged 39'

Lavin had lived at Warsop four years, during which he had been ill and practically confined to his lodgings.

E

CHAPTER 3

THE DATUM LINE STRIKE

The 'Mines for the Nation' Campaign fell flat. The MFGB therefore
returned to its more normal approach for increased wages. Now,
however, there was a different emphasis in the approach. In the days
prior to the 1914–1918 war, when, owing to the constant expansion
in the amount of coal demanded, coal prices rose faster than prices
in general, the miner looked to higher prices for increased wages.
Indeed, in many areas, formal sliding-scales which provided for
wages to move up and down with prices, had been in being for a
very long time. Further, at times when prices were rising (and,
therefore, demand was rising) the miner would be making full-time;
whereas when prices were falling, then wage-rates would be falling
and earnings would be falling still faster on account of short-time
working. Let us say, in short, that prices and wages moved up and
down together.

The post-war situation was quite different. Taking July 1914 as
100, the cost of living in June 1918 stood at 200; by January 1919 it
had moved up to 220; it fell back to 205 in June 1919 but by the
following January it was up to 225 and by June 1920 had reached
250. In this situation the tendency was for real wage rates to fall.
The average earnings per man-shift for the first quarter of 1920 at
15s 1½d were 233 per cent of the 1914 figure of 6s 5¾d.

By the beginning of 1920, the Federation was determined either to
secure a reduction in the price of coal, with a view to bringing down
the cost of living, or to secure for its members a substantial wage
increase. The Government made it quite clear that it was not disposed
to bring coal prices down. There had, it is true, been a reduction in
the price of house coal of 10s 0d a ton in December 1919, but the
price of industrial coal was not reduced. Since the Government were
not prepared to take any positive step to reduce the cost of living, the
MFGB pressed its claim for an increase in wages.

The claim was for an advance of 3s 0d a day flat rate (1s 6d for
those below sixteen) and the offer of the Government was for an
increase of 20 per cent on gross earnings (exclusive of 'War Wage'
and 'Sankey Wage') with a proviso that 'no person of eighteen
years or upwards shall receive a less advance than 2s 0d per shift or
per day worked, no person under eighteen years of age, but not
under sixteen years of age, a less advance than 1s 0d per shift worked

66

or per day worked, and no person under sixteen years of age a less advance than 9d per shift worked or per day worked'.[1] This increase, which was accepted after a ballot vote had been taken, had effect from March 12, 1920.

Within a few weeks of the signing of this agreement, the price of household coal was raised by 14s 2d a ton, and of industrial coal by 4s 2d a ton. This increase was seen by the leaders of the MFGB—and especially by the President, Bob Smillie—as an attempt to turn public opinion against the miner once more whilst, at the same time, facilitating the return of the industry to private enterprise. The MFGB estimated that the industry would make a net profit between June 1, 1920 and June 1, 1921 of £66,000,000, assuming no change in demand, wages or prices. Lloyd George, at a meeting held at the Board of Trade Offices on July 26, 1920 said: 'I am not going to dispute about that figure—I do not think it is very wide of the mark provided that we continue to export the same amount of coal and at the same prices.'[2] The importance of the export market at that time cannot be over-estimated. Whilst the average pithead price per ton to the domestic consumer was 33s 3d per ton, the pithead price per ton to the European consumer was 75s 0d, and Lloyd George made the point that since the domestic consumer already had this great advantage in price, it was illogical that he should be advantaged still further by the price reduction for which the MFGB were asking.

The claim now put forward by the MFGB was for an increase in wages of 2s 0d a shift and in addition, a reduction in the price of coal equal to the recent increase. Nottinghamshire, however, at the MFGB Conference which opened at Leamington on July 6, 1920, moved for an advance of 4s 0d a shift in wages and for the reference to the price of household coal to be dropped. This amendment was moved by Frank Varley and seconded by A. J. Cook, one of the South Wales militants, who was to become probably the most famous, and certainly the most controversial, General Secretary the Federation had ever had. George Spencer, however, supported Smillie on this question. At a Conference held on August 12, 1920 he said that the 14s 2d per ton increase in the price of coal was unnecessary, since the industry was already making a handsome profit, and that it was 'a vicious form of indirect taxation'. He argued further that if the union wished to win the support of the British public 'in this struggle, which has got to take place', it must stick to the policy of demanding a reduction in the price of coal. He went on to appeal to those leaders who were opposed to the majority view to drop their own individual opinions for the general good of the movement.

Mr Spencer's appeal was probably addressed to his colleagues in the Nottinghamshire coalfield, some of whom may have continued to campaign even after being defeated in Conference, for a 'straight' wage demand without reference to a price reduction. Be that as it may, when the membership was balloted on this issue, Nottinghamshire turned out to be very lukewarm. The question put to members, quite simply, was: were they prepared to come out on strike in support of a demand for an increase in wages and a reduction in price? In Nottinghamshire, 55·05 per cent of the men voted in favour of strike action compared with a national average in favour of 71·75 per cent. Only one County—Yorkshire—proved to be less enthusiastic than Nottinghamshire, with a percentage in favour of 51·11 per cent. At the other end of the scale, over 89 per cent of the Lancashire men voted in favour of coming out.

To some extent, the Federation claim for a reduction in price was probably designed to win the support of the other members of the Triple Alliance: railwaymen and transport workers. Every effort was made to keep the Alliance abreast of events.

A Conference of the Alliance was held on August 31, 1920. The President, Bob Smillie, in explaining the miners' case, pointed out that the bulk of the industry's current profits were made out of exports. Now, under the old Conciliation Board procedure, the wages of miners working in the exporting districts would have absorbed a substantial part of the increased revenue. However, the old arrangements had been brought to an end by the Federation because they felt that wages should be fixed on a national basis and the Government were appropriating the excess profits and using them to reduce the war debt. As Mr Vernon Hartshorn explained at the same conference, the owners were given a guaranteed profit of £26,000,000 plus 10 per cent of any excess over that figure. The other 90 per cent of that excess went to the Government.

The other two unions sympathized with the case put by the MFGB but J. H. Thomas, for the National Union of Railwaymen, put his finger upon a weak point. Referring to the impending miners' strike, he asked whether the strike funds would be shared equally by the district associations. Bob Smillie admitted in reply: 'If we enter this fight the districts who have been longest out of trouble and who have been saving the most money will be in the best position financially to fight. Our districts will not be equal as yours are.' He added that while some districts might be able to keep up strike pay for ten weeks, 'others might not be able to keep up strike allowances for more than three weeks'.[3] This very factor was to plague Nottingham upon a future occasion.

At a full delegate conference of the Triple Alliance held on September 22nd, Havelock Wilson, the seamen's leader, who was later to play an important role in the formation of 'non-political' miners' unions, expressed his doubts about the wisdom of strike action. 'I believe,' he said, 'that a strike would be disastrous to a good many organizations in this country.' He went on to argue that it was important to have the support of public opinion and that, therefore, there was 'room for a settlement of this dispute without any strike.'[4]

Whilst Havelock Wilson expressed the extreme view, there was clearly a reluctance on the part of many other delegates to rush into a sympathetic strike. J. H. Thomas moved that a deputation should be sent to see the Prime Minister to '. . . make another plea in the interests of the community as a whole as well as of the Triple Alliance and the miners'. This motion was seconded by George Spencer in the face of strong opposition from some of the other MFGB delegates who thought that it was wrong for one of their number to second a motion which had not first been considered by the MFGB delegation separately. The chairman ruled that Mr Spencer was in order, but it was 'clearly a matter that one would have liked to have avoided'. He went on to say that it was 'entirely a matter of taste'.[5]

On the day previous, September 21, 1920, George Spencer had similarly spoken in favour of a proposal which was repugnant to the majority of his colleagues. The Government had proposed in reply to the MFGB claim that the wage question should go to arbitration although they were not prepared to negotiate at all on the question of prices. The MFGB, on the other hand, were prepared at this stage to refer the question of prices to a tribunal but felt that the wage claim should be conceded 'forthwith'. Spencer pointed out that the Federation demand was illogical, and he suggested that both questions should be referred to arbitration. He thought that it was more important that prices should come down than that money wages should go up: '. . . it is very questionable,' he said, 'whether we are not in a worse position as to wages relating to prices than we were before we asked for the 20 per cent (advance in wages); therefore . . . I attach . . . more importance to the question of prices . . . than I do to the question of wages.' His attitude may, to some extent, have been due to the knowledge that, in Nottinghamshire, only 71 per cent of the men had handed in their notices in response to the union's instructions. Mr Varley might well say: 'You may be perfectly well assured of a sufficient number of men having given in notices to bring the collieries to a dead stand;' but even so, a degree of lukewarmness among the men was apparent.

To return to the question of whether or no a deputation from the Triple Alliance should be sent to see the Government in an endeavour to secure a peaceful settlement, the Triple Alliance Conference on September 22 agreed that the three organizations (miners, railwaymen and transport workers) should meet separately in order to mandate their delegates. At the MFGB meeting held early in the afternoon of the same day, George Spencer moved that the MFGB should agree to the proposal to send a deputation to see the Government. In doing so, he touched on the real weakness of the Alliance. The MFGB insisted that it alone should decide on what action to take in the event of a dispute, whilst at the same time expecting the other two unions to give full support. The other unions refused to act merely as adjuncts to the Miners Federation. If a decision affecting their members was to be taken, then they insisted upon having a say in the decision. The Triple Alliance could only be made to work properly by transferring to a joint body power to act in a dispute; but this the MFGB refused to acknowledge. Speaking at this same conference, Mr W. Carter, MP, referred to a speech made by Mr J. H. Thomas at the Triple Alliance Conference which Mr Carter took to mean that '. . . if the railwaymen and the transport workers entered into this business to assist, they would have to be in the negotiations'. Mr Smillie could not accept this interpretation of J. H. Thomas's speech: 'It would be a most amazing thing for them to be in at the negotiations when it was not their own case. I do not think Mr Thomas meant that . . . it would raise a very disgraceful scene if we met the Prime Minister and Mr Thomas felt he was there to negotiate for the miners.'

Despite misgivings, the MFGB Conference agreed to the sending of a deputation from the Triple Alliance to interview the Government 'to further urge the acceptance of the miners' claim'. The conference then adjourned.

The Triple Alliance deputation met the Prime Minister and his colleagues at 10, Downing Street, later in the afternoon. The Prime Minister suggested that the wages claim should be submitted to arbitration, or alternatively, that the owners and miners should agree upon an increase in wages tied to increased output. The idea of referring the wage claim to a tribunal was unlikely to gain acceptance. Indeed, at a previous meeting with the President of the Board of Trade, Sir Robert Horne, held on September 9, 1920, Mr Smillie had made a complaint which has an ominously modern ring. He said: 'I believe . . . that some time ago the Government itself, or some Government department, laid it down that there were to be no further increases of wages, that the vicious circle had gone far enough, and from that time down to the present there has not been any upward change of wages, I think, given by that Court.' He went on to ask: 'Do you think it is possible for persons drawn from the commercial and wealthy classes to be impartial on a question of this kind?' Because of the strength of feeling which existed on this issue among miners' leaders, Lloyd George must have known how unlikely it was that the proposal would be accepted. In putting the proposal forward once more, however, he saw clearly enough that an outright rejection of arbitration might help to turn public opinion against the miners.

The Triple Alliance Conference met once more, on the morning of Thursday, September 23rd, to hear a report from the deputation which had met the Prime Minister. Bob Smillie gave a short report after which the conference adjourned until 5.55 pm in order to allow the three organizations to consider the situation separately. At the Miners' Conference, Mr W. Hogg of Northumberland, moved that 'notices for the moment be suspended (they were due to expire two days later) and our people given a further opportunity of declaring what their will is in this matter' (of an arbitration tribunal). This was seconded by Mr E. Hough, of Yorkshire, and supported by George Spencer, who returned to his suggestion that the MFGB should propose to the Government arbitration on both wages and prices, and so 'place the responsibility for any strike on the Government . . . because they were not prepared to submit their case to arbitration'. The most powerful supporter of the proposal to accept an arbitration tribunal, however, was the President, Mr Bob Smillie. He said that, irrespective of his own private convictions, he had in negotiations opposed arbitration because it was his duty to put the majority view. However, he had 'thought for some time on this question differently from the majority of the Executive'. He was certain that the miners' case for increased wages was so strong that they would have no difficulty in convincing a tribunal that they were right. He was afraid that a strike would weaken and, to some extent, disrupt the Federation; and he believed that a strike would be used by the Government in a future election campaign, thus ensuring an increased right-wing majority and postponing the day when a Labour Government might be in power (and, therefore, be able to nationalize the mines).

Smillie was attacked by the militants of Lancashire and South Wales, whilst Herbert Smith, Vice-President of the MFGB, disowned the point of view of his fellow-Yorkshireman Mr Hough. Nottingham's William Carter, on the other hand, was proud of Smillie for his 'wise and sane and statesman-like speech'. He believed that if the Federation were to fail to take the opinion of the men, it would go 'headlong to destruction and disaster'. He pointed to the great division of opinion which existed and announced that, if a ballot vote were to be taken, he would advise his members to vote for arbitration.

When the issue was put to the vote, eleven districts, mustering 360 votes, voted in favour of referring the question of a tribunal back to the men, whilst nine districts, with 545 votes, voted against.

The delegates then attended the adjourned conference of the Triple

Alliance, where the President, Bob Smillie, explained that the MFGB had voted against suspending notices in order to take a ballot vote on the question of going to a tribunal; so that the miners' notices would terminate within forty-eight hours.

The spokesmen for the railwaymen and transport workers explained that they had not yet reached any final decisions, but that they would have difficulty in persuading their members to take sympathetic action since the miners had refused the Prime Minister's offer of arbitration, whilst both the NUR and the transport unions had themselves recently accepted arbitration for the settlement of their own grievances.

Later in the evening, after a further adjournment, the NUR spokesman said that his people had voted against coming out on strike. An acrimonious debate followed and Bob Smillie, from the chair, had to remind delegates that they could not 'afford to fight each other:' there would be enough fighting to do outside the Conference.

On the following day, Friday, September 24th, there were still further conferences. The Executive Committee met the Prime Minister once more in order to inform him that the miners would stop work on the following morning. The Prime Minister suggested that the Federation should suspend notices for a week, and that in the meantime, they should meet the owners and try to thrash out a scheme for increased wages dependent upon increased output. The Executive Committee agreed to recommend this course to the delegate conference. At this conference, Frank Hodges explained the position fully and said: 'After very careful deliberation the Executive came to this conclusion, that we recommend the Conference—our own Conference—to agree to a suspension of notices for one week, and let the Executive Committee be empowered . . . to meet the owners at once with a view to fixing up a settlement with them and report to a Conference to be held in London on Thursday next.' This suggestion was opposed by A. J. Cook, who accused the leaders of betraying their members: a suggestion which met with the well-merited scorn of Bob Smillie. Finally, the motion to suspend notices was carried by 134 votes to 31.

This decision had then to be communicated to the Triple Alliance Conference, where it was received with relief if not delight. The sole disturbing note at this session of that much-adjourned conference was sounded by Ernest Bevin who said that the Triple Alliance had 'revealed itself as a paper alliance this week'. He urged that some thought should be given to the reorganization of the Alliance so as to avoid future revelations of weakness. It was generally agreed that the Joint Committee should consider this matter, but in fact the weakness persisted.

The Conference ended with the Miners' Federation Executive sending telegrams to districts, advising them that notices 'were suspended for a week'. However, a week proved to be all too short to bridge the gap between the most that the owners were prepared to offer and the least that the men were prepared to accept. In brief, the union wanted an immediate increase of 2s 0d a shift, with further increases depending on increased output; whilst the owners insisted that any advance should be conditional on increased output. During the week, minor concessions were made by the two sides. The Executive reported on these discussions at a National Conference held on Thursday, September 30, 1920. At this conference, William Carter said that the Nottinghamshire Council had voted in favour of a further suspension of notices. This earned the amusement of the South Wales syndicalist, George Barker.

The chairman, Mr Smillie, said that they could do one of three things: to strike on Monday morning; to explore further the possibility of an output bonus; or to go before a tribunal. He was against a strike since he desired to find: '. . . before getting into a strike, whether or not we are likely to be successful in that strike. If this great organization fights nationally and fails I fear it will take many years; it would take, I think, the young men in this organization many years to build it up again to the present strength. If we fail, the mineowners would then secure the opportunity helped by the Government, who do not like and have not liked, the power of this Federation, would take the opportunity of attacking us sectionally.' He went on to urge once more the desirability of going before a tribunal.

This suggestion was strongly opposed by the South Wales militants. Mr S. O. Davies, for example, thought that the real strength of the Federation lay in the fact that '. . . since particularly 1916, capitalism in the mining industry is rapidly failing to function'. This notion that capitalism was on its last legs and that the Federation should help to put it into its grave by, for example, strike action, lies at the root of many of the differences between the left and right wings of the movement.

Mr George Barker, also of South Wales, declared: 'I am going to speak, for the few minutes I shall speak, unhesitatingly in favour of a strike.' He went on to characterize, by implication, those who were in favour of attempts to increase production as 'jackals of the capitalist class'.

George Spencer pointed to the position of the poor districts who had no more than £1 a member in the till, and he suggested that if they were to consider the idea of a strike seriously, they should pool their funds nationally. Spencer then turned to a favourite subject of his: the dropping of the demand for a reduction in coal prices, which he referred to as the 'greater question'. He went on to say: 'If the two claims had been as one, I would have been prepared to have a strike, but, having divided it into a wage claim, it is not worthy of a strike in view of the other. The Government refuse to grant the claim, and we say we are not going to a tribunal, then we are up against the Government. I think it is up to us to accept the challenge and go to a tribunal, seeing the other part of our claim is dropped. In my opinion we ought to ask the men to accept that suggestion.'

Finally, Conference agreed to adjourn in order to allow the Executive Committee to pursue the matter further. On the following morning (Friday, October 1st) the Executive Committee once more met the owners, this time with the Prime Minister present. Following this meeting, a further report was presented to the adjourned conference, when the delegates decided to suspend notices to Saturday, October 16th and that the owners' offer should be the subject of a ballot vote.

The owners' offer may be summarized as follows: If during the first fortnight of October the output is expected to reach the rate of 240 million tons a year, then an advance of 1s 0d a shift shall be paid. Should the output exceed the rate of 240 million tons a year, a further increase of 6d shall be paid for each additional 4 million tons.[6]

On Wednesday, October 13th, the MFGB Executive Committee met to receive the result of the ballot, which showed a huge majority against the offer. The Committee, therefore, decided that 'in view of the ballot vote the men be advised to allow the notices to expire, and that a cessation of work take place as and from Monday, October 18, 1920'. This recommendation of the Executive Committee was endorsed by a delegate conference held on the following day, October 14, 1920, although Mr W. Hogg of Northumberland, made one final attempt to have the matter referred to arbitration.

Mr Spencer then moved 'that the Executive Committee be instructed to interview the Prime Minister and the Cabinet and place before him the decision of this Conference, and ask him if he is now prepared to give the 2s 0d, 1s 0d, and 9d, and this Conference stand adjourned for that purpose and meet again tomorrow to hear the report of the Executive'. This proposal met with the vociferous opposition of the left wing delegates (plus Frank Hodges) and was defeated. Instead, it was decided that the Conference's decision should be conveyed to the Prime Minister in writing.

On the day following, Conference considered the Prime Minister's reply which 'made capital out of the advice some of us conscientiously gave to the men', as Smillie said. This letter, which regretted the men's decision to refuse arbitration, was written with one eye on the state of public opinion. It also probed the division existing inside the Federation between those who favoured arbitration and those who opposed it.

Smillie himself felt that the rejection of arbitration was a vote of 'no confidence' in himself, and he tendered his resignation. However, he was prevailed upon to continue in office although he did so with reluctance.

And so the miners entered upon this national strike a sadly divided body. A great many of the district leaders, and in particular, Hogg of Northumberland, Hough of Yorkshire, and Spencer of Nottingham felt, as Smillie did, that they should have accepted arbitration. Arnot characterizes the 1920 strike as 'a mistaken effort' because 'it drained the accumulated strike funds of the miners without yielding any lasting gain and created the precedent that an appeal to their partners in the Triple Alliance could be made without result'.[7] This view may be 'hindsight', but many of the more cautious leaders saw this point before the strike began. They believed also that it would be virtually impossible for a tribunal to flatly reject their claim having regard to all the circumstances.

Certainly the strike cost the NMA dear. Some £132,000, representing the union's net income for fifteen months, was paid out in strike pay.

The strike began on Saturday, October 16th. For all anyone could see, it was likely to last a very long time. However, on October 21st, a decision taken by the National Union of Railwaymen brought an unexpected accession of strength to the miners. This decision was conveyed to the MFGB Executive, at its meeting held on October 23rd, in a letter which reads as follows:[8]

'NATIONAL STRIKE OF MINERS

'Dear Sir,

'I am instructed to convey to you the following resolution carried at a Special Delegate Meeting at Unity House today:

"That this Special General Meeting having carefully considered the position created by the Miners' Strike, and being satisfied their claims are reasonable and just and should be conceded forthwith, decides to instruct the General Secretary to intimate to the Prime Minister that unless the Miners' claims are granted or negotiations resumed by Saturday, October 23rd, which result in a settlement, we shall be compelled to take the necessary steps to instruct our members in England, Scotland and Wales, to cease work."

'Yours faithfully,
'C. T. CRAMP.'

At the same meeting, a letter was read from the Prime Minister inviting the office-bearers of the Federation to meet him 'for the purpose of making an attempt to arrive at a basis of settlement'. Accordingly, on the following day, Sunday, October 24, 1920, the officials of the Federation met the Prime Minister and his colleagues at No. 10, Downing Street. Subsequently, on the following four days, negotiations with the Government continued. The Government's final offer has been summarized by Mr Page Arnot as follows:[9]

'First, the advance of 2s 0d a shift was conceded immediately and was equated to a figure of output. Secondly, additional increases of wages were set out on a scale corresponding to further increases of output. Thirdly, the scheme was to continue only till the setting up of a National Wage Board, to be negotiated upon between owners and workmen, who were bound to report on a permanent scheme not later than March 31, 1921. Fourthly, the Government were to guarantee the 2s 0d a shift advance to the end of December in any event: and were to guarantee export prices at 72s 0d a ton.'

As with the previous offers, this final offer sought to relate wages to a certain level of output; and because of this the strike has come to be known as the Datum Line Strike.

A Special Conference of the Federation was held on Wednesday, November 3rd, to receive the result of a ballot vote on the Government's offer. Sixteen districts had voted in favour of acceptance by majorities ranging from 244 (Kent) to 28,428 (Yorkshire). Four districts had majorities against: Lancashire with 55,509, South Wales with 46,495, the small district of Forest of Dean with 196, and Nottinghamshire with 915. It is true that 915 is quite a small majority in a total poll of over 28,000, but to find Nottinghamshire in the same camp as the three notoriously left-wing districts is surprising, particularly when the NMA Council 'strongly recommended the members to vote in favour of accepting the terms as submitted by the Executive':[10] a recommendation which was conveyed to the men at meetings held at various central points in the County. It should be borne in mind, however, that 16,000 of Nottingham's 44,000 members did not bother to vote.

Because of the heavy majorities of Lancashire and South Wales, there was an overall majority of 8,459 against acceptance of the offer. The President explained, however, that the rules of the Federation provided that: 'If a ballot vote be taken during the time a strike is in progress a vote of two-thirds of those taking part in the ballot shall be necessary to continue the strike.' This ruling met with the strenuous opposition of the militants, particularly Greenall of Lancashire, and

Cook, S. O. Davies and Ablett of South Wales. One speaker from South Wales, Mr H. Jenkins, pointed out however, that by no means all Welshmen supported the rejection of the Government's offer. He said: 'I venture to submit the intelligence of the men in my district against the men following men like Cook.' He also expressed a view similar to that of Herbert Smith who, despite his disapproval of arbitration, said: 'Districts are talking about going back to the old position. If you do not want the 2s 0d, well and good. I shall advise Yorkshiremen to have it, you do as you like.'

The chairman's ruling was eventually upheld by the Conference, 121 delegates voting in favour whilst forty-six voted against. By a rather narrower margin (103 votes to forty-six) Conference decided to advise the men to return to work on the following day, November 4, 1920.

And so the Datum Line strike was over. But its consequences were still to make themselves felt. The strike coincided with the downswing of the Trade Cycle following the post-war replacement boom. During the summer months of 1920, the level of employment in certain consumer goods trades moved downwards. The coal strike of 1920, despite the large stocks which had been accumulated beforehand, 'resulted in general slackness and much unemployment and short-time working in nearly all the principal industries'.[11] Thereafter, following a partial recovery when the strike was over, the level of employment continued to fall. The percentage of unemployed among members of trade unions making returns rose as follows:

1920		
April	0·9	
May	1·1	
June	1·2	
July	1·4	
August	1·6	
September	2·2	
October	5·3	(Datum Line Strike occurred here)
November	3·7	
December	6·0	
1921		
January	7·1	
February	8·7	
March	10·2	

True, the downturn occurred earlier than the Coal Strike; but the strike undoubtedly played its part in speeding the development of the depression.

It appears that the slackening off in industrial activity started in the leather and textile industries which had benefited from the 'buying spree' which followed the war-time period of shortages. The depression spread gradually to other consumer industries, and then came the Datum Line strike, with its effect on transport and heavy industry; and more important still, its effect on the outlook of businessmen.

In its early stages a slump is largely a psychological condition. Manufacturers and traders sense a slackening off in industrial expansion, which makes them cautious: they delay capital projects and allow stocks to run down a little. Lenders on mortgage become uneasy and they start to call in their money. People who have borrowed on mortgage find themselves in difficulties and, if they are lucky, liquidate their debts by selling stocks at knock-down prices; if they are unlucky, go bankrupt.

The coal strike of 1920 was fought out against such a backcloth. The Government, which could have helped matters, was too much concerned with the New York exchange to worry overmuch at the prospect of falling prices in Britain. And so it allowed the Bank rate to remain at 7 per cent throughout the winter, when prices were collapsing. According to the figures prepared by the Federal Reserve Board, wholesale prices in Britain fell in this period as follows:

	All Commodities
1920	(1913 = 100)[12]
April	334
May	340
June	339
July	326
August	322
September	315
October	297
November	280
December	260
1921	
January	244
February	226
March	213

As Professor Pigou says: '. . . the collapse of prices led, as it was bound to do, to an attack, among other wages, upon those of coal miners, and so was ultimately responsible for the disastrous coal strike of April-June 1921, which, of course, in turn inflicted heavy damage on industry.'

The Datum Line strike did incalculable harm, not only to the miners, but also to the working-class in general. Further, it was the worst kind of mistake: a mistake that could have been avoided; a mistake that would have been avoided had all the leaders read the signs of the times correctly, as Smillie did.

The ironical part of this is that, just as Smillie in his younger days had attacked the then leaders of the Federation, so now Smillie was attacked by A. J. Cook and his kind; and so was Cook in his turn to be attacked, at a later date, by Arthur Horner.

CHAPTER 4

THE 1921 LOCKOUT

1. *The Events Leading to the Lockout*

Following upon the patched-up peace of November 1920, the Coalowners' and Miners' organizations sought to arrive at a permanent settlement. Under the terms of the temporary settlement, the two sides had until March 31, 1921 to 'report on a permanent scheme'. However, little progress was made.

At the same time there was a very wide measure of agreement between the two sides on fundamental principles: both sides agreed that there should be a consolidation of wage-rates in place of the hotch-potch of percentages, flat-rates and bonuses which had grown up; both sides agreed to the formation of output committees; and both sides agreed that owners and workmen should share the proceeds of the industry in agreed proportions. Had the economic climate been milder, it is conceivable that an amicable settlement would have been arrived at. The future of the industry might then have been far different from what it turned out to be.

Unfortunately, the economic climate was not mild. The boom of 1919–20, which lasted little longer than a year, was followed by a period of depression without parallel in British industrial history. In 1921, coal owners and miners alike were victims of Government policy. The Government was determined to 'make the pound look the dollar in the face'. This policy entailed the depression of domestic price levels, and since labour costs bulk so large in the price of coal, the miners were bound to suffer more than most.

To make matters worse, the Government felt that 1921 was the appropriate time to return the mines to private ownership and control. After all, the days of huge profits wrung from continental customers were now almost over and the mines were likely to prove a liability rather than an asset in the years ahead.

In their early stages, the discussions between the MFGB and the owners went well. At a meeting of the Federation Executive Committee held on December 9, 1920, it was resolved:

> 'That districts that have not already done so, be urged to now set up district output committees, and give effect to the resolution of the Joint Committee of owners' and workmen's representatives.'

The NMA Council at its meeting of December 28th endorsed this

MINERS

OF

NOTTS. & DERBY.

You are asked to be **LOYAL** to the Federation !

You **HAVE BEEN LOYAL**
at the expense of your savings, your wages, your domestic happiness.

WHAT IS YOUR REWARD ?

Debts, empty hopeless promises, a few shillings here and there, Russian money DRAGGED from desperately poor Soviet workers ! and now

A COUNCIL OF WAR !

Against Whom ?

Against What ?

AGAINST THE INDUSTRY !
YOUR JOBS !
YOUR BEST INTERESTS !

BE MEN, NOT CHILDREN, and FACE THE FACTS !

Crease & Mugglestone, Printers, Station St. Mansfield.

2. The coalowners' answer to A. J. Cook's campaign to bring out again the men who had returned to work during the 1926 stoppage

resolution and decided to join with the owners in setting up output committees at pit level in addition to the District Committee. However, the pit committees did not get into their stride, possibly because of the declining demand for coal which made them unnecessary. At any event, the local owners did not evince much interest in the idea.

There was, however, one fundamental issue dividing the two sides: the Federation wished to continue the practice of national settlement of wages and a national wages pool, whilst the owners were determined to return to the pre-war practice of district settlements. The miners' attitude can be expressed quite easily: 'We will not, if we can avoid it, I assure you,' said Robert Smillie, 'see the wages of the miners in one district of this country being reduced down to the starvation point or below starvation, while miners in other parts of the country more favourably situated are receiving fair remuneration for their labour.'[1] Further, from the point of view of the left-wing, to abandon the national wages pool would be to take one further step away from the ideal of social ownership and control.

The owners' view can be explained just as easily: to make efficient coalowners subsidize their less efficient colleagues through a National Wages Pool would act as a brake on efficiency. It would help high-cost producers to stay in business using resources (especially skilled labour) which could be used more effectively elsewhere. Further, by evening-out differences in wage-rates it would penalize the more hard-working and more efficient miner, thereby reducing his incentive to continued hard work and efficiency. Further, the owners also saw that to retain a national wages pool would be to strengthen the hands of the nationalizers.

To add to the difficulty, the Government made it clear in the early months of 1921 that they proposed to return the mines to private ownership and control. In a booklet issued for the Mining Association, this process was described briefly thus:

'In March 1921, the Government was losing £5¼ million a month on the industry; and, as the then Chancellor of the Exchequer stated at the time, in arriving at this figure nothing was allowed for any profits to the owners, nothing for depreciation, and nothing for interest on borrowed money. This loss—at the rate of £63 million a year—had of course to be made good by the taxpayer and so alarmed the Government that the industry was hastily decontrolled.'[2]

Of course, the industry was not decontrolled quite so hastily as all that. An intimation of the Government's intention was given to the MFGB Executive by the President of the Board of Trade at a meeting

F

held on January 5, 1921. It was not until February 23rd, however, that the Federation learned, in discussions with Government representatives, that complete decontrol was to take effect on March 31st, and not, as both miners and owners had hitherto believed, on August 31st at the earliest. The Federation took the Government's decision as a breach of faith, as the following passage shows:

> 'MR ROBERT SMILLIE: I suppose we may take it that the passing of the control, if it passes at the 31st March, means that the Government have no further responsibility in connection with wages or in connection with the profits of the employers.
> 'THE PRESIDENT (OF THE BOARD OF TRADE): That would be so.
> 'MR ROBERT SMILLIE: I think you are aware that the employers' view of the matter is that that would be a breach of a pledge given to them when the Act was passed.
> 'THE PRESIDENT: So they have informed me; but I have done my best to disabuse them of that view.
> 'MR ROBERT SMILLIE: Well, we are of opinion that it is a breach of an understanding so far as we are concerned. . . .'[3]

Frank Hodges expressed the view that: 'If you decontrol when we cannot sell our coal except at a very low price you have only a couple of districts or so that can continue remunerative production, namely, Nottinghamshire and Derby, and perhaps Yorkshire;' whilst Frank Varley pointed out that Cumberland, to take an example, was working at a loss of 14s a ton, and he asked how the men could possibly be expected to bear this out of their wages.

It will be appreciated that, during the period of control, whilst the 'actual management of individual mines was left to their respective owners' the Mines Department directed them 'to dispose of their supplies so as best to meet the needs of the country'.[4] Wages were negotiated nationally and were paid for out of a national fund; whilst a proportion of excess profits (after 1920 the whole of excess profits) was paid into a fund from which the owners of unprofitable mines received compensation along the lines of a cartel.

Decontrol would end all that. The various wage increases granted by the Coal Controller during the period of control would be taken off whilst the high-cost pits would no longer receive a subsidy to help them keep open. The position of marginal producers in the absence of national control is well illustrated by the following extract from a speech by Mr Evan Williams, President of the Mining Association, at a meeting of the two sides on April 26, 1923:

> 'In those districts like Cumberland, Lancashire and North Wales, it is a matter of wonder, I think, how the owners have been able to

keep going for so long, making losses as they have been. It is true collieries are difficult things to close down; they are kept working at a loss rather than incur a greater loss by closing them; there is always the hope in the minds of both owners and workmen that better times will come; that the sacrifice they are making will cease to have to be made, and that prosperity will return to the district after all.'

Again, at a similar meeting on November 15, 1923, Mr Williams said:

'. . . but I believe the truth of the matter is that there is a point where loss in working coincides with the loss in stopping, and until that point is reached collieries will continue to go on. The loss is less. But that cannot go on for ever; and there is no doubt that financial accommodation has had to be made for these people, over and over again, in the hope that times would improve and that some of these losses might be made good again: I think it is, in the case of a large number of collieries, the expectation of profit upon which they live and not any realization of an average.'

'MR STEPHEN WALSH (MFGB): It is that hope which springs eternal in the human breast.'

'MR EVAN WILLIAMS: Yes, that is it.'

The owners would have preferred to put off the evil day for a time, but like the miners, they were the victims of Government policy.

The Miners' Federation was faced now with the possibility that wages would be slashed all over the country, and that many of its members would become unemployed through the closure of un-economic pits, thus intensifying still further the tendency for wages to fall.

On March 24, 1921, the Coal Mines (Decontrol) Act became law. By this time also, all the miners in the country had been served with notices to terminate their contracts whilst the various District Coalowners' Associations had offered new terms to their District Miners' Associations. In some districts, the proposed reductions were so steep as to invite rejection out of hand. Even in Nottinghamshire, where wages need not have been reduced at all, the Owners' Association were asking for a substantial reduction to meet 'contingencies'.[5] However, the NMA could not negotiate on the terms since the official policy of the Federation was to insist on a National Wages Pool.

On March 17th the Executive Committee of the MFGB met to consider the proposed scheme for the future regulation of wages submitted by the Mining Association. The owners' document begins:

'It being agreed that wages in the industry must depend upon the financial ability to pay, the owners propose that the following principles be adopted by the Mining Association of Great Britain and the Miners' Federation of Great Britain for application to the determination of the *wages payable in each district* upon the *financial position of such district*.'[6]

The principles can be summarized thus:

1. Basis rates, plus the percentages ruling in July 1914, to be regarded as the minimum wage. Piece-workers shall also retain the addition (14·2 per cent in the case of Nottinghamshire) paid consequent upon the reduction in hours from eight to seven a day.

2. Owners' standard profits to be calculated as 17 per cent of the aggregate wages payable as in No. 1 above.

3. Any surplus left after the standard wage, standard profits, and other costs have been met shall be shared between owners and workmen—25 per cent to the former and 75 per cent to the latter. The workmen's share (if any) to be expressed as a percentage addition to basis rates.[7]

It appeared from the figures presented to the MFGB Executive Committee that the reductions in gross wages would range from 1·03 per cent in Derbyshire, Nottinghamshire and Leicestershire to 47·55 per cent in Cumberland and Westmorland; the average reduction for the country being of the order of 22 per cent.

Faced with a situation as serious as this, the Committee advanced for the consideration of districts, the tentative proposal that the policy of a national wages board and national pool should be temporarily shelved; in order that district associations might be free to secure the best terms possible for their members.

At a meeting of the NMA Council, held on March 22nd, it was resolved 'That this Council cannot agree to the abandonment of the claim for a National Wages Board even temporarily, it therefore refers the whole matter back to the Federation Executive Committee to further urge the acceptance of our proposals for the unification of the industry for a national pool, and for the owners to receive a uniform 10 per cent allocation of the aggregate wages. It further demands a joint effort to be made with other large organized industries for united resistance to the general movement for reductions in wages.' This was moved by Lowmoor and seconded by Newstead. A further resolution—moved by Huthwaite and seconded by Summit—asked the Executive Committee of the MFGB 'to consider the most practical means of working the pits in the interests of the operatives' in the event of a lockout.

These two motions, coming as they did from left-wing branches, give us an indication of the programme of the syndicalists: a refusal to abandon the National pool (despite the fact that in the long run such an abandonment would almost certainly be in Nottingham's interests); a general strike to prevent wage reductions; and, to follow this policy to its logical conclusion, the taking over of the industry by the Federation.

However, despite the resolution of Council, George Spencer took quite a different line. At a special National Conference held on March 18th when the EC's motion asking districts to take the views of their members on the desirability of abandoning the idea of a national wages board and pool temporarily came up for consideration, Spencer moved that the words 'temporary' and 'temporarily' should be taken out of the motion. He explained that he thought that the EC should press as vigorously as possible for the national board and national pool. He then went on to argue that: 'if we relinquish the idea of a national board, it is not going to be temporary, but we shall do so in the consciousness that it is going to be permanent until we have really nationalization of the industry.'

Spencer's view was that of a realist: he recognized that there could be no half-way house in the conditions of the 1920s. If they were to go back to the old system of independent control and district settlements, then they could not expect to regain a national pool until, as he said, 'there is a fundamental change in the Constitution'. However, when his amendment was put to the vote, only five delegates voted for it. Conference decided, by a large majority and in the face of strong opposition from the militants, to refer the question to districts. The moderate leaders, like W. Hogg of Northumberland, saw the dangers in insisting on the full programme: to fight on this issue would dissipate the union's resources and would allow 'international competitors to get our markets' which we might have difficulty in winning back.

S. O. Davies of South Wales, on the other hand, wanted to 'throw down the gauntlet' in order to bring private ownership to an end: an utterly impractical programme in the circumstances of the time.

Conference adjourned until March 24th to give districts an opportunity to vote on the issue. As we have seen, Nottinghamshire voted against the idea of opening district negotiations and abandoning temporarily the national agreement and pool. So, too, did twelve other districts with an aggregate voting strength (including Nottinghamshire) of 727; whilst seven districts, with a voting strength of 241 votes, voted in favour. The only large districts to support the recommendation were Yorkshire and Northumberland.

At its meeting on March 30th the Council of the NMA decided that all men except those engaged on winding up horses should cease work on Thursday, March 31st when their notices expired. Those engaged on winding up horses were to be allowed to work until Friday night, April 1st, when they too must cease.

This was in line with the decision of the Executive of the MFGB—reached on the same day—that all men (including pumpmen, ostlers, enginemen and firemen) should cease work at March 31st. This decision was a sad mistake; since it put into the hands of the Government and the owners a propaganda weapon to turn the public against the miners. The picture of the 'poor dumb' pit ponies left underground without food and water, though an untrue one, served its turn.[8] More to the point, the calling out of the pumpmen in many cases caused flooding which rendered some pits unusable for months, thereby leaving a legacy of unemployment long after the conclusion of the dispute. (To his credit, Herbert Smith admitted the error in his report on the dispute.)[9]

The Executive Committee then interrupted its sitting to go to meet the President of the Board of Trade, Sir Robert Horne, in order to report to him that the negotiations with the owners had reached a deadlock, and to ask whether the Government could give the industry any further financial assistance to enable the employers to pay a living wage. Mr Hodges was now the Federation's principal spokesman, Bob Smillie having resigned the presidency on health grounds.

Sir Robert rejected outright the suggestion of a subsidy, and he then asked whether the Federation had scrutinized the offers made by the district associations of coalowners. Mr Hodges said that they had, and various members of the Committee then proceeded to say what they thought of these offers. Mr Varley complained that the owners of Nottinghamshire and Derbyshire had departed from the general principles for calculating wages laid down by the Mining Association. The February results would give, using the Mining Associations' principles, a percentage on basis rates of 141, but the owners had decided 'to keep back, to meet contingencies, 31 per cent' giving 'an additional deduction of 2s 5d a day' making a total deduction of 2s 6d instead of 1d.

Unfortunately for Nottinghamshire, the policy of the Federation on a National Agreement precluded negotiations with the owners' associations on the basis of their offer. A local agreement on the pattern of the Leen Valley settlement of 1893 was ruled out.

2. *The* 1921 *Lockout*

On Friday, April 1st, the pit wheels were at a standstill. However, neither the owners nor the Government seemed particularly anxious to get them going again. Indeed, they refused to meet the MFGB unless the latter would first instruct the 'safety-men' to go back to work so as to avoid permanent damage to the workings.

The Government did, however, meet the Triple Alliance partners of the miners (the National Union of Railwaymen and the Transport Workers' Federation) on Saturday, April 9th, and again insisted that negotiations with the miners could not be opened until the safety of the pits was assured. Accordingly, at the behest of their allies, the MFGB reluctantly agreed to let the 'safety-men' go back to work, and instructions were issued to districts. Following this decision, the Government opened negotiations with the Federation, but little progress was made. On the side of the Federation, there was undoubtedly an expectation of a change in the Government's attitude following the sympathetic strike which the Railwaymen and Transport Workers had called for Friday, April 15th, and whilst the Federation offered to accept a reduction of 2s 0d a shift in wages, in the light of the fall in the cost-of-living, they were not prepared to give up the principle of a national agreement and pool.

Then on April 14th, Frank Hodges addressed a meeting at the House of Commons following which, in answer to a question, he was understood to say that the MFGB would be prepared to abandon temporarily the National pool idea. This was seized upon by the Prime Minister, who invited Hodges 'and his fellow delegates to meet the owners at the Board of Trade . . . to consider the best method of examining the question of wages'. This letter was considered by the MFGB Executive who replied that: '. . . the only condition upon which a temporary settlement can be arrived at is one that must follow the concession of the two principles already made known to you, viz. a National Wages Board and National Pool.'[10]

Whatever Hodges may actually have said, this contretemps gave the NUR and the TWF the excuse they needed to call off the sympathetic strike, and the miners were then left to fight it out alone.

In the event, it was to be a war of attrition. The demand for coal at this time of year tends to be on the low side anyway, and in 1921 the Government were not too unhappy at the prospect of deepening the depression which was expected to restore the sacred pound sterling to its throne. A fair amount of foreign coal was coming into the country—despite the opposition of the continental miners' and dockers' unions—because as Mr Hodges reported to the MFGB Annual Conference on August 17, 1921 '. . . Mr Havelock Wilson's men (i.e. the Seamen's Union) decided that at all costs they would load the coal in face of the fact that the Continental dock labourers refused to load it'. In Nottinghamshire, all the pits were picketed; a committee representative of the NMA, the Deputies' Union and the Enginemen and Firemen's Union, being responsible for the arrangements. The pickets had to face a certain amount of provocation from the police, and in some cases they had to face proceedings under the Emergency Powers Act. On May 30th Council protested against the action of the Government 'in prosecuting members of the community for exercising the right of free speech'. Four of the union's members at Selston were sentenced under the Act, whilst Joseph Webster of Ilkeston was summoned for assaulting a reporter at a meeting.

As the lockout lengthened out, the NMA Council took steps to strengthen its hand. On April 25th it decided to call out all men except those engaged on providing water for local inhabitants and this instruction was confirmed at subsequent meetings. A week later, Council resolved that no coal should be wound out of any pit during the continuance of the dispute. This was a clear attempt to prevent the owners from working the engines even if the enginemen (most of whom belonged to the Enginemen and Firemen's Union) presented themselves for work. This policy would appear to run contrary to the revised instruction of the MFGB designed to assure the safety of the pits.

Council also attempted to stop its members from producing coal from outcrops and dirt tips: an extremely profitable business at a time when coal was being sold at famine prices. This decision emphasizes the fact that the dispute was not merely with the owners (who received no benefit from the sale of outcrop coal), but with the Government. Had the union been in dispute with the owners alone, then surely it would have been sensible to organize outcrop working along business lines, using the proceeds to augment the strike funds? However, the Federation aimed to bring the economy to a standstill and thus force the Government's hand. There is here a clear contrast with the 1893 lockout, where the sole enemy was the Coalowners' Federation; and where the Government acted as independent arbiters.

Council also attempted to stop the Digby Company from disposing of slack from their Gedling pit bank.

Towards the end of April, the Government put forward a definite offer of a subsidy: a grant of £10 million to help the industry meet its wage bill. Sir Robert Horne, now Chancellor of the Exchequer, explained that:

'. . . We propose that in the month of May no greater reduction should take place in the wage of the miner in any district than 3s 0d a shift, and in the month of June no greater reduction should take place than 3s 6d a shift. The remainder of the sum of £10,000,000 to which I have referred should thereafter be divided between the months of July and August, two-thirds in the month of July and one-third in the month of August. . . .'

Sir Robert went on to say that a uniform reduction in wages was not necessarily involved; although if the MFGB wished the reduction to be uniform over the whole country, the Government would be willing to make the necessary arrangements. He pointed out, however, that in this event, in districts like South Yorkshire, Leicestershire, Nottinghamshire and Derbyshire, 'the miner would be taking less wages than the mineowner is willing to give him'. The Government's offer was, however, conditional on a durable settlement being made. Sir Robert went on to mention the one big obstacle standing in the way of a 'durable settlement': the Federation's insistence on a national pooling of profits. The Government regarded this as a political question requiring legislation; and it was not prepared to introduce legislation for this purpose; and it was for the people at a General Election to say whether the Government was right or wrong.

Nevertheless, whilst the Government could not agree to the idea of a national pool, it considered a national wages board as essential. Whilst actual terms would be settled on a district basis, the national board would lay down the general principles by which the district organizations were to be guided.

This offer was considered on the afternoon of the same day (April 28, 1921) at a national conference of the Federation, when it was turned down by 890,000 votes to 42,000, Northumberland being the only district to vote in favour. At a meeting of the NMA Council, held two days later, the national conference decision was accepted, but it was agreed that delegates should explain the Government's terms to their members.

On May 27th, the Executive Committee of the Federation met the Prime Minister and his colleagues, together with the Central Committee of the Mining Association. The Prime Minister was careful to point out that very little pressure had been brought to bear on him to arrange a settlement. During the war and the boom period which followed it, '. . . every day [lost] meant a great loss to the community-owners, manufacturers, agents, middlemen as well as workmen . . . everybody said, "Here we are losing our time; we are losing our harvest". But it is so no longer, and I am going to speak quite plainly. There is no doubt that some inexperienced men, with a very superficial knowledge of the conditions of business and of the economic conditions of the world came to the conclusion that they were dealing with the same situation now; that they had the same means at their disposal for bringing pressure; that the community would at once say, "For God's sake, settle, and pay anything". But the situation is completely different, and no man is more alive to it than I am.' Lloyd George was undoubtedly right in ascribing to the 'militants' of the MFGB an unwillingness to face facts. 'Direct action' at a time of slump could only make sense if the organized working class were prepared to seize power; and this, they were manifestly unwilling to do. For all that, the forces of law and order were taking no chances. Not only were crowds of policemen drafted into mining areas, but many reservists, including some members of the NMA (e.g. Joseph Lievers, Walter Deakin, Walter Booth, Willie Smith and Henry Green), were called to the colours in case of civil commotion. Herbert Smith, in his reply to the Prime Minister, protested at great length about the proceedings brought against members of the Federation under the Emergency Powers Act and at the many instances of police provocation. He spoke about a meeting which had been held at Barnsley, 'where the policemen batoned innocent people' and argued that incidents of this kind would 'cause more rebellion than there has already been in this business'.

The offer made by Lloyd George on this occasion was little different from the offer made in April. Again the Government would allot £10 million to allow the reductions in wages to be applied gradually; and again, district settlements were envisaged.

This offer was submitted to districts who were asked to report their views to the General Secretary by Friday, June 3rd. The Nottinghamshire Council resolved: 'that our representative on the Executive be instructed to stand for the National Pool, and the members recommended to re-affirm their previous decision on the question'. When the National Executive met on June 3rd, the Secretary reported that all the other districts had taken the same decision as Nottingham. This decision was conveyed to the Prime Minister, who intimated in reply that the Government had nothing further to offer, and that the offer of the £10 million would only remain open for another fortnight. This warning precipitated a further approach from the Mining Association who invited the Executive Committee of the MFGB to meet them.

This meeting took place on Tuesday, June 6th, when the owners presented their revised offer; and on the following day, the Executive decided to convene a national conference and, meantime, to put in hand arrangements for a ballot vote. The ballot paper asked:

'Are you in favour of fighting on for the principles of the National Wages Board and National Pool, with loss of Government subsidy of ten million pounds for wages if no settlement by June 18, 1921? ....

'Are you in favour of accepting the Government's and owners' terms as set forth on the back of this ballot paper? ....'

On the back of the paper, the terms of the government and owners' offer were summarized. The Government were offering 'ten million pounds to prevent large reductions in wages' where reductions were necessary; the immediate reduction was not to exceed 2s 0d a day, and there were to be no further reductions until August 1, 1921. In addition, there was to be a permanent scheme to operate when the Government grant was exhausted. This provided for a National Wages Board 'to fix principles for guidance of districts'; to fix the ratio of profits to wages; and to fix the amount of the new standard wage. The new standard wage was to equal the total wage paid in July 1914 (plus district additions to standards), plus the percentage for pieceworkers caused by the reduction of hours from eight to seven, and a minimum percentage of 20 per cent added thereto. The board was to be composed of equal numbers from the two sides, with an independent chairman, and there were to be similarly constituted district boards.

The owners put out a series of tables to show the effect of their proposals. The rate to be paid in Nottinghamshire was shown to depend upon the grouping to be adopted. If the areas covered by the Nottingham and Erewash Valley Owners' Association (i.e. mainly Leen Valley pits) and the Midland Counties Owners' Association (i.e. mainly Erewash Valley pits in Derbyshire and Nottinghamshire) were to be regarded as separate districts, then the average shift rates under the permanent scheme would be 9s 7¼d and 9s 3¾d respectively. Indeed, it appeared that the Leen Valley men might receive a small increase in wages instead of a reduction. If, however, these areas were to be grouped along with the rest of Derbyshire, Leicestershire, Cannock, Warwickshire and Yorkshire the average would be 8s 10¼d. (A table at Appendix A summarizes the position.)

The National Conference of June 10th considered these proposals and confirmed the decision to ballot the men on the owners' offer. Conference also decided that the men should not be given any recommendation as to how to vote.

The Executive Committee of the Federation met to receive the result of the ballot vote on June 17, 1921. Only two 'districts'—the Cokemen and Enginemen (occupational groups affiliated to the MFGB in the same way as the district miners' associations) were in favour of accepting the Government and owners' terms. Among the miners' district unions, however, there were wide variations in the degree of support for the 'Fight for principle' (of a National pool).

Of Nottinghamshire's 46,206 members only 15,069 troubled to vote: less than one-third! Of those who voted, 8,099 voted for the 'Fight for principle' and 6,970 voted for settling the dispute. The table in Appendix B to this Chapter shows the voting figures for the various districts. The 'unenthusiastic' ones; those, that is, with a very small majority for continuing the struggle, were the ones that had little to lose from the proposed terms of settlement: Derbyshire, Leicester, Midlands, Nottingham and South Derby, plus Northumberland, which did stand to suffer a substantial reduction but where labour relations were traditionally good.

South Wales, Scotland, Lancashire and Forest of Dean, the 'militant' districts, on the other hand, had a great deal to lose under the proposed terms of settlement.

So far as Nottingham is concerned, as we have seen, only one member out of three troubled to vote at all (which puts this vital issue at the level of a Rural District Council Election) and only one in six voted in favour of continuing the 'fight for the principle' of a national pool. The fact is that the Nottingham miner had little at stake, as in the 1912 Minimum Wage Dispute; and whilst principles are all very fine, they butter no parsnips.

The result of the ballot vote was conveyed to the Prime Minister who, after expressing regret for the Federation's decision, emphasized that the Government's 'offer of assistance cannot remain open after tomorrow night'.

The Executive Committee, at its meeting on June 18th, discussed the situation at great length and resolved:

'That we ask various Executive Committees of unions affected by wage disputes to meet the Executive Committee of the Federation with the object in view of taking national action with the miners to secure their mutual demands.'

In view of their lack of success with the Triple Alliance, the older members of the Executive Committee at least must have realized that this hare would not run.

In addition, some member of the Executive (presumably either Noah Ablett or A. J. Cook) gave the following notice of motion for the next meeting:

'That the Executive Committee decide to withdraw the safety men.'

However, it was obvious by now that reality had to be faced: supporting action by other unions could not be relied on; such desperate measures as withdrawing safety men (even if the safety men obeyed, which is not at all certain—many of them in Nottinghamshire ignored Council's instructions to cease work) were not likely to bring agreement nearer; and in the meantime, suffering was intense.

In Nottinghamshire the funds of the union were completely drained, and in addition the Association had plunged deeply into debt. Now more than ever, the officials had cause to regret the ill-conceived stoppage of the previous year.

The following sums were expended on lockout pay in this period:

	£	s	d
Five weeks ended April 23, 1921	92,409	19	7
Five weeks ended May 28, 1921	53,026	6	2
Four weeks ended June 25, 1921	40	6	2
Four weeks ended July 23, 1921	7,940	7	0
	£153,416	18	11

Source: Accounts in NMA 1921 *Minute Book.*

In addition, the Association borrowed money wherever it could, and issued coupons to members when hard cash ran short. These coupons were, in effect, promissory notes to be used to purchase goods from local traders, and they bore interest at the rate of 6 per cent per annum although one firm, J. D. Marsden & Co, would not accept payment of interest but held the coupons as an interest-free investment. In June, members were being paid half the normal strike pay in coupons.

However, if things looked black in Nottinghamshire, they looked still blacker in those parts of the country where, as Arnot says: 'for many a day the miners' wives and families had not had enough to eat, and were able to keep going in many cases only by the soup from the communal kitchens set up by the Lodge Committees'.

Consequently, despite the result of the ballot vote, the Executive Committee of the Federation decided, on June 24th, to 'ask the Government and the owners for a meeting with a view to negotiating a satisfactory wages agreement which we can recommend our members to accept'.

The meeting was held on June 27th, after which the Executive Committee recommended districts to accept terms which were substantially the same as those which the membership had so recently rejected.

The Executive Committee's recommendation was conveyed to the NMA Council at its meeting on June 29, 1921. The minutes of this meeting read:

'1. Moved by Mansfield, seconded by Bulwell—That Mr Varley be thanked for his report.

2. Moved by Lowmoor, seconded by Bentinck—That we recommend our members to accept the report of the Executive and resume work.'

When the Executive of the MFGB met on Friday, July 1st to receive the district vote, it learnt that some thirteen districts had voted in favour of accepting the proposed terms of settlement, whilst five districts (Lancashire with 90,000 votes, and four small districts: Bristol, Forest of Dean, Kent and Somerset with 15,820 votes between them) voted against. An immediate return to work was therefore ordered.

This decision had been anticipated by Lancashire who put in 'an emphatic protest against the action of the officials and Executive Committee of the MFGB in ignoring the decision of the rank and file in the recent ballot vote, and over-riding the authority conferred upon them by abandoning the National Wages Board and the National Pool of profits and wages . . .' and more to the same effect. Lancashire demanded that a special National Conference should be called 'so that the whole situation can be submitted to the judgment of the rank and file'. This request was turned down.

Arnot ends his account of the lockout:

'The lockout was ended.

'The miners were defeated. They had come through great tribulation and the end of it was not in sight. But they were not dismayed, for they had in them the knowledge that they would live to fight another day.'[11]

For the present writer the lockout has a different significance. The Executive might argue that 'Up to now the unity of the men has been magnificent: whole districts which had nothing to gain in the form of wages have stood loyally by the other districts whose wage reductions would have been of the most drastic character'; but this unity and loyalty had been very badly strained. In Nottinghamshire, the vociferous minority of idealists might give the appearance of solidarity, but behind this appearance was the reality of an apathetic majority. Many of those who did not take the trouble to vote in the ballot to decide whether to continue or end the stoppage, were busily at work on the dirt-tips and outcrops. And many of them were to drop out of the union in the months ahead when they found themselves called upon to pay stiff levies to pay off the mountain of debt accumulated during the lockout.

It is not difficult to visualize the conclusions which men like Spencer and Varley drew from the lockout. Here was a dispute where Nottingham stood to gain nothing, In defending a principle (which involved the permanent subsidization of the poorer counties by the richer, with Nottinghamshire as the chief loser from the arrangement) their members had suffered hardship and their Association had been financially crippled. And further, as we shall see in our next section, the owners had used this opportunity to reverse the trend towards daywork and to force the men back to filling by screens and forks instead of shovels. And this sacrifice had achieved precisely nothing. Is it not reasonable to suppose that these men would expect, in future, to have the interests of Nottinghamshire considered as well as the interests of the larger (and poorer) districts?

The 1921 lockout could afford satisfaction only to those who believed that every fight is a useful rehearsal for the final struggle leading to the overthrow of capitalism. For those who believed that it was the union's job, within the framework of the existing system, to extract the largest possible share of the proceeds of the industry for their members; this lockout, like the strike of the previous year, was a catastrophic waste of effort and money.

There were many estimates of the loss caused by the lockout, and of these, this estimate of direct losses given by the Labour Correspondent of the *Nottingham Guardian* seems as good as any:

LOSS OF INCOMES DIRECTLY CAUSED BY 1921 LOCKOUT[12]

Miners' Wage loss at £4 million a week	=	£52 million
Loss of Wages in Iron, Steel and allied trades at £5 million a week	=	£65 million
Loss of Wages in Consumer goods industries and distributive trades at £5 million a week	=	£65 million
Plus: Additional Expenditure by the State	=	£75 million
		£257 million

The losses in miners' wages continued for some time after the dispute was over since many of the collieries could not be restarted immediately. At some pits (e.g. Newcastle, one seam at Broxtowe and Nos. 7 and 9 pits at Langton) the workings were waterlogged and took a great deal of pumping dry. At most pits the fires had been drawn and it took a few days to get up steam, set the pumps working and put the engines in order. At other pits, only datallers could be employed during the first week or so owing to the state of the shafts or roadways.

The Labour Correspondent of the *Nottingham Guardian* expressed a common view when he said:

'There is only one redeeming feature in this disastrous struggle, but it is a feature which all the parties will be able to look back upon with pride and no little satisfaction. That is the good temper, the good behaviour, and the good order which has prevailed from the beginning to the end of the conflict.'

Many of the men had spent the time in their gardens; others had taken charge of the arrangements for feeding children in local schoolrooms and for raising funds with which to buy the food; whilst in the Hucknall and Annesley district there was hardly a church or a chapel which had not been redecorated by voluntary labour during the lockout.

The main exception to the general rule of peaceful behaviour involved—in 1921 as in 1893 and 1844—the Watnall Colliery of the Barber, Walker Company. The official 'history' of the Company[13] gives the following account of the incident:

'Members of the clerical staff rallied round by firing boilers and carrying out many jobs to which they were unaccustomed. Mr Phillips and Mr Stafford drove locomotives. These activities brought no remonstrance or demonstration from the Company's workmen. Indeed, some of them commenced organized opencast working, the sales of coal from which were managed by a check-weighman [Mr Joseph Birkin], an action which indicated the men's indifference to the strike, for it not only helped to retain a local supply of coal, meagre though it was, but it also helped to 'break' the strike.

'In due time these happenings at Eastwood Collieries were noised abroad, for smoke issuing from Watnall Colliery chimney had been detected by a few rampant Union men at Hucknall and Bulwell. They organized others of that ilk, and on June 23rd three mobs approached Watnall by different routes, with the avowed intention of quenching the boiler fires. The Company's Police Constable, Bonnet, was alive to the situation, and through his action and that of company officials, the County Police were informed that a disturbance of the peace was imminent. Before the mobs arrived, two bus loads of County Police had arrived at Watnall in charge of a wide-awake Inspector. Their conveyances were hidden from view in the yard of Brooksbreasting Farm, and the Police were disposed all out of sight at strategic points near the boilers. The approach to the boiler-firing platform was only about four feet wide between the high brick walls of the boiler housing. The mobs attacked in force, their ringleaders deservedly forming the van-guard. The pressure of their lusty colleagues behind forced the ringleaders and many others through the narrow aperture into the arms of the waiting constables who belaboured the attackers with truncheons and, pulling them in, passed them on to other waiting constables. There were many broken heads by the time those behind realized what was happening. When they fled, leaving the Police to render first-aid to the casualties, a few of whom were taken to hospital. Thereafter quiet prevailed at Watnall.'

It is clear from the above account that the Police precipitated a breach of the peace instead of preventing it (which they could have done by guarding the pit gates), and this reflects no credit on them. A second breach of the peace was narrowly averted when a crowd of men from the Hucknall district went to Eastwood with what Mr Whitelock calls the 'avowed intention of lynching Birkin'. Mr Birkin, with a bodyguard of two policemen, met the crowd near the 'Horse and Groom' Inn, Moor Green, where he was met with a storm of abuse. After a few minutes, apparently, he persuaded his detractors to talk things over with him in the 'Show Field'. Mr Whitelock ends his account: 'Ere long Birkin had offered to argue the strike situation before them all, at the Sun Inn croft, and there they repaired to receive the lesson which awaited them.'

Whatever Mr Whitelock may feel about it, Mr Birkin was an official of the Nottinghamshire Miners' Association who was quite deliberately acting contrary to the instructions of his union. His attitude was disloyal in the extreme. However, there can be no doubt that this likeable demagogue exercised considerable power over his members at Eastwood. Long before 1926, the Nottinghamshire Miners' Association Branch at Moor Green had assumed the shape of a company union; and Mr Birkin, as we shall see, supported an organization which advocated 'non-political' trade unions. If any one man can claim to be the father of the Spencer union it is, I think, Birkin rather than Spencer himself.

Much of Birkin's influence prior to 1926 was exercised through a 'Central Committee' consisting of representatives of the Moor Green, High Park, Brinsley, Underwood, and Lodge contractors. This unofficial body was frowned upon by the permanent officials of the Association who (with the exception of J. G. Hancock) were anxious to bring the 'butty' system to an end. The 'Central Committee' usurped to some extent the normal functions of the union, and its existence was to facilitate the organization of a breakaway in 1926.

Another example of unruly behaviour occurred at Brierley Hill Colliery in April. Here, four men threw stones at officials who were keeping the pumps and boilers going.

3. *The Day-wage System and the Shovel Filling Question*
The Nottinghamshire Miners' Association had succeeded, in 1919, in forcing the owners to allow stallmen to fill coal by shovel and not by screen or fork. They had also succeeded in introducing either the

G

day-work system or, more frequently, the all-throw-in system, throughout the County.

However, these victories had, from the outset, an appearance of impermanence. Various owners attempted to re-introduce screens, or go back to the 'butty' system; and it was obvious that the union could expect trouble over these issues when the economic climate changed.

The leaders of the NMA felt very strongly about the day-wage system. Charlie Bunfield, speaking at a conference of the MFGB, pleaded powerfully for a national campaign to abolish piece-work in the interests of safety. 'I do not know whether you have it in any other counties,' he said, 'but we have had, rightly or wrongly, a colliery in our district (presumably Cinder Hill) which for the last two years has been on the day-wage system; 9s 5d a day plus everything else which has come along. How many fatal accidents do you think there have been at this colliery during the last two years? Not one. 'I know you may say to me: "Yes; but it is a coincidence." It may be coincidence, but the fact is there.' Mr Bunfield argued that it is no good earning £8 one week and death the next, and he suggested that the Federation should give fourteen days' notice to terminate contracts on this issue.

On February 2, 1921, at a meeting between the NMA and the owners, the union applied for all workers in normal stalls to be paid a flat rate of 8s 3d a shift but nothing appears to have come of this. Indeed, the time for securing reforms of this kind was past: the owners were now intent on reducing labour costs which they believed could best be achieved by piece-work systems of payment; and by paying only for large coal.

The coalowners of Nottinghamshire had always been keen on the quality of coal turned, and a favourite device for ensuring a profitable product was to insist that coal should be filled by screen or fork, so that all small coal became 'waste'. This device had the effect of reducing earnings considerably since much of the coal got had to be left in the pit. Frank Hodges, writing in 1920, gave the following estimates of small coal wasted in this way:

Nottinghamshire & Derbyshire	574,000 tons per annum
South Derbyshire & Leicestershire	184,000 tons per annum
Warwickshire	65,000 tons per annum
South Wales & Monmouth	1,502,000 tons per annum

Hodges went on to say: 'Small coal is cast back into the mines in other districts, although not to the same extent . . .'[14]

During the boom of 1919–20, the men at nearly all pits refused to use screens; but gradually the screens were brought back again at many places. This led to the following resolution adopted at a meeting of the NMA Council held on February 28, 1921:

'That this Council strongly protest against the introduction of screens and forks into the mines of this County, and insist that where they are now in use in any mine that they should be withdrawn and that we urge the workmen in their own interests to fill the coal as clean and as free from dirt as it is at all possible for them to do.'

It was decided also, that a copy of this resolution should be sent to Mr C. H. Heathcote of the Sherwood Colliery Company.

On April 25th, when the lockout was in its fourth week, the shovel-filling question was again discussed by Council when it was resolved: 'That it be an instruction to all our members that they do not use in any seam or pit, for the purpose of filling coal, either fork or screen.' However, the owners used the occasion of the lockout when distress was widespread and the men were eager to get back as quickly as possible, to insist upon the re-introduction of screens when the pits resumed work. This was done, for example, by the Staveley Colliery Company, Derbyshire, and other owners in the two counties quickly followed suit: competition tends to make this kind of action infectious.

During August a meeting with the owners took place when the latter urged on the Association the necessity for re-introducing screens in view of the difficult trading conditions with which they were faced. Council agreed that Branches should be asked for their views on this. As was to be expected, the suggestion met with an unfavourable response.

On October 11th, the Midland Counties Coal Owners' Association gave notice to their workmen in Nottinghamshire and Derbyshire that on and after October 26th, shovel filling would not be allowed, but screens would have to be used instead. Mr Frank Hall, Secretary of the Derbyshire Miners' Association argued, in a letter to the EC of the MFGB, that:

'. . . the owners under the National Settlement are unable to insist on this proposed change insomuch as it would alter the wages, and would mean a very considerable reduction in the wages of workmen whose earnings depend upon the amount of mineral gotten.'

A conference was held between representatives of the Derbyshire

and Nottinghamshire Miners' Associations and the Midland Counties Coalowners' Association, but the owners refused to accept the unions' views on this question. The Nottinghamshire Council therefore '. . . empowered the officials to negotiate with the coalowners in an endeavour to arrange terms for the retention of the practice of shovel filling in the pits of the county'.[15]

Apparently, the owners were unwilling to treat on this question: at any rate no further mention is made of it in the NMA *Minute Book*. Agitation continued at various pits however. At the Bolsover Company's pits a joint committee, with Tom Pembleton and German Abbott of Rufford as its leading spirits eventually succeeded in forcing the owners to agree to the use of shovels.

Some owners were also determined to re-introduce the 'butty' system, and they were aided and abetted by a section of the workmen themselves.

The agreement under which the all-throw-in system was introduced was negotiated jointly with the Midland Counties' Colliery Owners' Association by the Derbyshire and Nottinghamshire Miners' Associations, so that, when the system came under fire, the two unions consulted together on the question. In December 1921, a conference was held at Derby. George Spencer reported on this meeting to Council on December 31, 1921, when it was resolved that: '. . . we recommend the members to continue to throw-in, and share equally in the earnings, and that the officials have power to take into Court any case where the owners have reduced the member's wages from 8s 3d (i.e. the County Basis "make-up" rate for fillers) to any less sum.'

The men were now faced with the alternatives of yielding to the pressure of their employers or following the recommendation of their union. The extent to which the all-throw-in system was adopted depended mainly on the determination of the men, and for this reason, the Association held propaganda meetings at Mansfield and other places designed to strengthen the determination of their members.

The mere fact that the officials of the union had to address public meetings in order to induce the men to follow union policy is itself an indication of weakness. Mr J. W. F. Rowe records the opinion of the union officials in 1923, that the agreement to work the all-throw-in system was being kept in the breach rather than in the observance.[16] And this in spite of Council resolutions, branch decisions and public meetings. There was, of course, a vested interest to be overcome within the ranks of the men. On this issue at least, the Association's boast, 'United we stand', rang hollow.

APPENDIX 'A'

AVERAGE EARNINGS PER SHIFT AT THE STANDARD (PRESENT BASE RATES AND PERCENTAGES PAID ON THEM IN 1914 OR THEIR EQUIVALENTS, AND THE PERCENTAGES GRANTED TO PIECEWORKERS IN 1919 IN CONSEQUENCE OF THE REDUCTION OF HOURS) WITH THE ADDITION OF 20 PER CENT THEREON, BASED UPON THE OUTPUT ACTUALLY OBTAINED IN MARCH 1921

District (*Arrangement 'A'*)	s	d	District (*Arrangement 'B'*)	s	d
Scotland	8	8½	Scotland	8	8½
Northumberland	8	1	Northumberland	8	1
Durham	8	9	Durham	8	9
South Wales	9	6	South Wales	9	6
Yorks, Notts, Derbyshire, Leices, Cannock & Warwick	8	10¼	South Yorks	9	1
			West Yorks	8	6¼
Lancs & N. Staffs	7	6½	East Yorks	8	7¾
North Wales	7	9¾	Lancs & Cheshire	7	9¼
S. Staffs & Salop	7	10¼	North Wales	7	9¾
Cumberland	8	11	Midland Counties	9	3¾
Bristol	7	9	Notts & Erewash	9	7¼
Forest of Dean	7	6¼	Other Derby	9	4¼
Somerset	7	5½	South Derby	9	4
			Leicestershire	7	11¼
			North Staffs	6	9¼
			South Staffs	7	5¾
			Cannock	7	8¾
			Salop	9	0¼
			Warwick	7	8¼
			Cumberland	8	11
			Bristol	7	9
			Forest of Dean	7	6¼
			Somerset	7	5½
			Yorks	8	11

Reprinted in MFGB *Minute Book*, 1921, p. 363.

APPENDIX 'B'

1921 DISPUTE: RESULT OF BALLOT HELD IN JUNE
(from Table on p. 391 of MFGB *Minute Book*, 1921)

District	For Continuing Fight for Principle	For Govt. and Owners' Terms	For Continuing Fight for Principle	For Govt. and Owners' Terms	Affiliated Membership*	Percentage Voting (Cols 1+2 ÷ Col 5)
Bristol	1338	224	85·66	14·34	2200	71
Cokemen	1593	3412	31·83	68·17	9800	51·07
Cumberland	5168	1548	76·95	23·05	13500	49·75
Derbyshire	11050	9948	52·63	47·37	49000	42·85
Durham	69991	20744	77·14	22·86	126240	71·87
Enginemen	3876	10411	27·13	72·87	20200	70·70
Forest of Dean	5222	659	88·79	11·22	6000	98·02
Kent	1208	272	81·62	18·38	2000	74
Lancashire	64084	7417	89·63	10·37	90000	79·45
Leicester	2537	2042	55·28	44·72	7900	58·09
Midlands	20030	15866	56·33	43·67	64000	56·08
Northumberland	14695	12758	53·53	46·47	41500	66·15
N. Wales	6474	3082	67·75	32·25	16300	58·62
Nottingham	8099	6970	53·74	46·26	35000	43·05
Somerset	3843	740	83·85	16·15	5620	81·54
S. Derby	1832	1612	53·19	46·81	6400	53·81
S. Wales	110616	40909	73·00	27·00	200000	75·76
Yorkshire	52829	28044	65·32	34·68	142500	56·76
Scotland	51129	14056	78·44	21·56	110000	59·17

* *Note:* The affiliated membership figure does not necessarily equal the actual membership of any district. Nottinghamshire, for example, had an actual membership of 46,206 at this date.

For the sake of comparison, the table below illustrates the effect of the owners' offer on average wage rates. This is extracted from a table prepared by the Mines Department and should be treated (as to the actual figures) with a certain amount of caution. The table does, however, show clearly the ordinal relationship (e.g. first, second, third, etc.) between the districts in the matter of wage reductions which could be expected to follow a settlement along the lines of the owners' offer.

REDUCTION ENTAILED BY OWNERS' OFFER

District	Reduction per shift* (Taken from Table in which no Adjustment for Stores has been made)		Reduction per shift† (Taken from Table where Adjustment for Stores has been made)	
	s	d	s	d
Scotland	5	8½	4	10¼
Northumberland	7	6½	6	2
Durham	3	9¼	3	4¾
S. Wales	11	7¼	9	3¾
Lancs & Cheshire	5	1	4	3½
N. Wales	6	10	6	6
N. Staffs	6	5¼	5	6¼
S. Staffs	6	7¼	6	4¾
Salop	8	2½	6	6½
Cumberland	1	11½	8	6½
Bristol	9	0	5	0
Forest of Dean	7	0	7	0
Somerset	3	11½	3	4
South Yorks	1	3½		10
West Yorks	3	8	3	1
Midland Counties (i.e. Notts & Derbyshire)	1	8		
Notts & Erewash		3 (increase)		9¾
Other Derby	1	2¼		
South Derby	2	5¾	2	3
Leicestershire		3 (increase)		8 (increase)
Cannock	2	7½	1	7
Warwick	1	11½	1	10

* *Source:* Table from Mines Dept headed: 'Cost to the State of Subvention, Owners' offer being taken before adjustment for stores, etc.'

† Ditto—but *after* adjustment for stores, etc.—MFGB *Minute Book*, 1921, p. 364.

Whilst there is no exact correlation, the following comparisons are significant:

A. *Districts which stood to lose more than 5s a shift* (*under the first arrangement*).

Name of Districts	Percentage of Affiliated Membership Voting	Percentage for Continuing the 'Fight for Principle'
Scotland	59·17	78·44
Northumberland	66·15	53·53
S. Wales	75·76	73·00
Lancs & Cheshire	79·45	89·63
N. Wales	58·62	67·75
N. Staffs	} No separate voting figures	
S. Staffs		
Shropshire		
Bristol	71·00	85·66
Forest of Dean	98·00	88·79

B. *Districts which stood to gain or to lose less than 2s a shift.*

Cumberland	49·75	76·95
S. Yorkshire	No separate voting figures	
Notts & Derbyshire }	*Nottinghamshire*	
Notts & Erewash	43·05	53·74
(i.e. Leen Valley)	*Derbyshire*	
Other Derby	42·85	52·63
Leicestershire	58·09	55·28
Warwickshire	No separate voting figures	

CHAPTER 5

THE AFTERMATH

1. *Unemployment*

The increase in unemployment following the collapse of the 1919–20 boom made itself felt in the mining industry early in 1921.

A special conference of the Miners' Federation of Great Britain held on Wednesday, January 26, 1921 and the succeeding days, adopted a resolution on unemployment which has a distinctly Keynesian flavour, as this extract shows:

'The Conference declares that, in a period of unemployment, the policy of the Government should be one of expansion, not of contraction, of rightful and economical public expenditure; and that the necessary public works and services which must certainly be executed within each decade, ought to be, as far as possible, concentrated on the years of industrial depression; so as to avoid the waste of keeping workers in one year in idleness upon unemployment benefit, and in another year, on excessive hours of labour at overtime rates.'

This resolution fell on deaf ears (on deaf Labour ears as well as deaf Conservative ears: where fiscal policy was concerned Phillip Snowden was as much a conservative as Winston Churchill); so that we are still, in this year of Grace one-thousand, nine-hundred and sixty-two, short of the motor-roads which could have been built in the inter-war years.

Speaking on this question, A. J. Cook advocated 'drastic action' to deal with unemployment. 'We may not solve the problem by a general strike,' he declared, '. . . but we shall have brought it to a head, then will come the situation where we as organized Labour have to tell the democracy whether we can take over the industries.' He was seconding a motion moved by his fellow South Welshman J. Winstone, which read:

'That we recommend, as the Government have refused to put into operation the policy of the Labour Party for dealing with unemployment, measures be taken to get the whole Labour Movement to take drastic action within fourteen days to enforce its policy.'

In reply, George Spencer said, speaking of the working class, 'Unemployment, more or less, is one of the sad consequences of our own folly at the General Election, and the people will have to learn from experience, a very sad experience of this character.' He went on to say that the widespread unemployment was due to under-consumption, and he thought that a deputation representative of the whole Labour Movement should urge upon the Prime Minister the desirability of increasing unemployment benefit so as to increase 'the volume of trade'.

Bob Smillie said bluntly: 'I have urged direct action again and again during the last few years. I have urged it at a time when I felt sure that the needs of the nation itself would make it successful if we tried. I am not of opinion that the present time is the time for direct action being taken. I believe absolutely that the employers would welcome it. . . . If we mean by a general strike we are going to throw over capitalism as we know it and institute a new system of Communism or ownership of industry, then we ought to come forward and say so, but we ought not to drift into a general stoppage on the question of forcing the Government out by political action and then allow ourselves to be led on the other hand to try to establish a new form of Government.' Later, he said that: 'I believe that capitalism never was stronger than at the present time; it was never so well entrenched and never so anxious in its history to create a stand-up fight with organized labour.'

Here, the master strategist was explaining the art of generalship to the militants of South Wales: if you fight you should be sure that you have a chance of winning. Spencer, on the other hand, put his faith in strictly constitutional action: the people must be left to put things right through the ballot box.

The Federation had aready faced one costly dispute (the 1920 Datum Line strike) and was shortly to enter another (the 1921 Lockout); to strike over unemployment was, therefore, out of the question.

The 1921 Lockout solved the problem of unemployment temporarily so far as the miners themselves were concerned; but it very greatly intensified the problem for other people. The percentage of unemployed (from the records of trade unions making returns) rose from 10 per cent in March 1921 to 23·1 per cent in June. The ending of the dispute brought an improvement—to 16·7 per cent—in July. Trade disputes in basic industries are bound to cause suffering and loss to people in other trades; and for this reason the interests of the public in general should, one feels, be considered before action is taken. The disputes of 1920 and 1921, in intensifying the depression as they did very substantially, caused widespread distress not merely whilst they were in progress but later, too. And, of course, the miners themselves were affected by this.

Unemployment pay in this period strained the resources of the NMA. Once the ex-servicemen had been absorbed into the industry— say by the end of 1919—out-of-work pay rarely cost the Association more than a hundred pounds a month until after the lockout. In January 1921 for example, expenditure on this account was £38; this went up to £3,630 in February because the eight hundred New Hucknall men were thrown out of work by a dispute between their employers and the railway companies. For March the expenditure was £1,604 and for April £263.

At the conclusion of the dispute, out-of-work pay was far more costly, as this table indicates:

OUT-OF-WORK PAY

Month		Amount (to nearest £)
August	(5 weeks)	£11,815
September	(4 weeks)	£12,422
October	(4 weeks)	£7,808
November	(4 weeks)	£4,833
December	(4 weeks)	£4,158

Source: Accounts in 1921 NMA *Minute Book.*

So serious had this drain become (the Association was, of course, very heavily in debt at this time) by January 1922, that it was decided that 'out-of-work pay be stopped in each and any case after thirteen weeks have been paid'. The union also had to reduce the old age pension from nine shillings a week to six shillings, and one branch, Mansfield, got into trouble for continuing to pay at the old rate.

During the summer months of 1922, Cotes Park worked only 15½ days in eight weeks, and they applied for a week's out-of-work pay but this was refused. This was an extreme case, but short-time working was, by this time, the lot of every pit. The market men in particular were suffering; making only odd shifts and half-shifts at intervals. The County Magistrates were so impressed by the distress in the Nottinghamshire mining villages that they communicated with the Prime Minister and put out an appeal to Boards of Guardians. As it happens, the fortunes of the industry were shortly to be improved by the French occupation of the Ruhr, which reduced foreign competition and thus led to increased exports, but in the meantime, the wives and families of Nottinghamshire miners found themselves faced with a second hungry Autumn.

2. The Soft Coal Agreement

Early in 1922 the local coalowners informed the Association that they could no longer keep their soft coal pits open at the existing rates of wages.

It will be remembered that, in the 1912 Minimum Wage Award, Judge Stanger divided the Nottinghamshire pits into Top Hard and Other Than Top Hard, and awarded higher rates for the former than for the latter. The owners had succeeded in convincing him that Top Hard coal was more profitable than any other, and that it could afford to bear higher wage rates. This did not affect wage rates normally paid however, since most people were enjoying higher basis rates than the legal minima.

Subsequently, the Mackenzie Awards of December 9, 1916 and September 5, 1918; and the Stoker Award of June 16, 1920, had widened still further the gap between the legal minima and wages actually being paid. The basis rate for haulage hands fixed by the Stoker Award for example, was 6s 8d compared with the legal minimum rate of 5s for Top Hard pits and 4s 10d for others. Now, however, the owners insisted that wage rates actually paid should reflect the difference in profitability between Top Hard and non-Top Hard pits.

The first meeting on this subject was held on February 2, 1922 at Derby, where the owners demanded a reduction in the basis rates of non-Top Hard pits of 10 per cent. They presented to the Association's officials evidence to show that losses were being made, as they might well have been at that particular juncture.

A further meeting was held on February 13th, when the Association spokesmen informed the owners that their terms were unacceptable. Further discussions followed and it was eventually agreed that basis rates for non-Top Hard collieries should be reduced by 7½ per cent. On March 27th, Council authorized its representatives to sign the Agreement, which came into effect on May 8, 1922.

During the negotiations on the revision of the Soft Coal rates, the owners had undertaken to lend their assistance in the matter of recruiting non-members into the union. What the Association had taken to be a firm undertaking turned out in practice to be nothing of the sort. In response to a request that the understanding should be honoured, the coalowners replied as follows:

'May 3rd, 1922.

'Dear Sir,

'Coal Seams Other Than Top Hard

'At a meeting of the above Coal Owners held this day in the Victoria Station Hotel, Nottingham, it was unanimously resolved that all Colliery Owners parties to the Agreement undertake to give all reasonable assistance with regard to non-union men that the majority of Colliery Owners have given in the past.

'Yours faithfully,

'F. CHAMBERS,
'*Chairman of Meeting.*

'F. B. Varley, Esq.'

On August 26th Council returned to the charge by demanding: 'That the officials seek an interview with the Coal Owners for the purpose of requesting them to enforce their promise re non-unionism, and that each local Secretary send on to these offices the attempts they have made with their management and the answers given on the same question.' Nothing came of this.

At this time, non-unionism was causing real anxiety for the first time in years. The growth in the labour force during the war had brought the Association's membership up to 46,000 and more in 1921, but after the lockout a sudden contraction took place. This was a perfectly natural reaction to the suffering caused by the dispute, and to the heavy levies (as much as 3s a week) then required of members to pay off the union's debts. Also, of course, many men were disgruntled with the terms of settlement and there had been a substantial minority opposed to acceptance of the terms. A union delegate, questioned by the *Nottingham Guardian* on July 1st, 1921, said:

'It does not promise well for the future of the industry that the men will return to work dissatisfied with the terms of the offer. . . . It is not over the disappearance of the pool they feel so sore. It is the likelihood of a heavy drop in wages after September that has stiffened their resistance.'

At four collieries, Gedling, Bestwood, Wollaton and Sherwood, the men voted against returning to work; and at most other places a substantial minority felt that they had been hard done by. There was, therefore, every reason to expect disillusionment to lead to a fall in membership. In August 1921, there were still (despite defections during the dispute) 45,684 members of whom 43,901 had paid all their dues. In August 1922, the membership total was down to 39,084, of whom only 23,764 had paid all their dues.

Most of the old established branches had a high proportion of 'financial' members, whilst the new, predominently left-wing led branches of the Mansfield Area had a low proportion. This can be seen from the table below:

OLD BRANCHES

Branch	Membership Total	Financial Members
Clifton	1400	1400
Annesley	907	807
Cinder Hill	895	817
Bestwood	1884	1774
Bulwell	469	447
Broxtowe	760	673

NEW BRANCHES

New Hucknall	982	290
Summit	2798	1278
Bentinck	1914	762
Mansfield	2099	1043
Rufford	920	200
Welbeck	1202	702
Lowmoor	928	216
Clipstone	120	50

Source: NMA *Minutes* 26/8/1922.

Of the 'old' branches, only Gedling (with 524 financial members out of 2934) and Moor Green (with 203 out of 833) had really bad records in this respect; and it is significant that at these two places the 'butty' system was most strongly entrenched. In any case, Gedling was not particularly old but it belongs geographically to the old part of the coalfield.

The Association organized trade union rallies to win men back to the union, particularly in the Mansfield and Eastwood areas, but the real obstacle standing in the way of an improvement was the very high rate of subscription demanded. Accordingly, on February 24, 1923, Council agreed to reduce the total contribution to 1s 6d per week, of which, 6d would go to pay the levy. This brought a slight improvement in 'financial' membership although total membership continued to fall. In February 1923 there were 35,313 members of whom 23,043 had paid all their dues, whilst in December there were 34,687 of whom 24,910 were 'financial'.

In our next chapter we shall see how this campaign for increased membership was intensified.

In the meantime Frank Varley, the Financial Secretary, improved the Union's administration. He chivvied branch secretaries and treasurers whose accounts were not in order, he made branches re-pay monies paid out in excess of the amounts specified in the rules, and he initiated reductions in the pay of permanent and local officials. Even so, the Association's debts exercised a crippling effect on its activities. George Spencer, speaking at a Federation conference on Tuesday, July 18, 1922, explained that the Association still had £90,000 to pay off. For this reason, he opposed the suggestion made by a number of districts that the minimum contribution for membership of a district union should be reduced from 1s 0d a week to 6d. South Wales had already, in the previous year, so reduced its rate of contribution, contrary to rule, in an endeavour to win members back to the union; but, as Spencer argued, the knowledge that other areas were doing things like that would be bound to have a bad psychological effect in Nottinghamshire.

In September 1923, when trade was beginning to revive, the Association applied for the restoration of the $7\frac{1}{2}$ per cent cut in non-Top Hard Wage Rates. However, the owners seemed in no hurry to reply to the claim, and on November 24th Council entered a strong protest. This brought the following reply from the owners:

'MIDLAND COUNTIES COLLIERY OWNERS' ASSOCIATION

'Dear Sir,

'Wages Agreement—Soft Coal Collieries

'Your application for restoration by Colliery Owners concerned of the reduction accepted under the above-mentioned Agreement was duly placed before a Meeting of those Colliery Owners interested, and I was directed to inform you that after due consideration the conclusion was arrived at that the application could not be entertained, the position not having materially altered since the Agreement was arrived at.

'Under these circumstances you will probably agree that no useful purpose could be served by the Conference which you suggested.

'Yours faithfully,

'WILLIAM SAUNDERS,
'*Secretary.*

'G. A. Spencer, Esq.'

The question was next referred to the Joint Board of Miners' and Enginemen and Firemen's Unions, where it was decided to tender fourteen days' notice to terminate contracts over this issue. However, this decision was never implemented which is hardly surprising considering the parlous state of the Association's finances, and the $7\frac{1}{2}$ per cent reduction was still being enforced at vesting day.

3. *The Eastern District*
We saw in Section 2 of our last chapter that the actual effect of any

settlement on Nottinghamshire would depend upon the grouping of
the districts. If the areas covered by the various owners' associations
were to be separate wages districts, then the Leen Valley would be
no worse off, and possibly better off than under the old arrangement
of national negotiations. This follows from the fact that, if wages are
to be governed by the ability of individual districts to pay them; then
the most profitable district will pay the highest wage. If, on the
other hand, a number of areas of varying profitability are to be
lumped together then the wages paid are bound to be lower than the
best area could afford to pay, and higher than the worst area could
afford to pay. Nottinghamshire in general and the Leen Valley in
particular therefore stood to lose by being included in a wages district
with other areas.

The offer made by the owners, which brought the Lockout to a
close, involved the creation of an Eastern District consisting of the
following Counties:

Yorkshire, Cannock-Chase, Warwickshire, Derbyshire, Nottinghamshire, Leicestershire and South Derbyshire.

Because of its higher basis rates, Nottinghamshire continued to be
the highest paid district. The hard fact is however, that a definite—
and avoidable—cut in wages was sustained as a result of the creation
of the Eastern District.

The following table illustrates first, the extent of the reduction, and
second, the comparative position of Nottinghamshire and its Eastern
District partners:

WAGE RATES IN THE EASTERN DISTRICT:
COLLIERS ON DAY WAGE:
MARCH 1921 COMPARED WITH NOVEMBER 1921

District	November Percentage on Basis	Basis Wage		March Rates		November Rates		Reduction	
		s	d	s	d	s	d	s	d
Yorkshire	110·55	7	6	17	11	15	9	2	2
Cannock	110·55	6	6	16	6	13	8	2	10
Warwick	110·55	7	0	17	3	14	8	2	7
Derbyshire	110·55	7	9	18	3	16	3	1	6
Notts	110·55	8	3	19	0	17	4	1	8
Leicester	110·55	7	0	17	2	14	9	2	5
S. Derby	110·55	6	6	16	6	13	8	2	10

Source: MFGB *Minute Book*, Table at p. 516. The information conveyed in this table appears to be correct for Colliers, though
inaccurate in the case of the other grades.

The method of determining wages now adopted was known as 'the ascertainment'. Each district had its own Wages Board and independent accountants. The standard wages were defined as 'the district basis rates existing on March 21, 1921, plus the district percentages payable in July 1914 . . . plus in the case of pieceworkers, the percentage additions which were made consequent upon the reduction of hours from eight to seven'. Standard profits were to be 17 per cent of the standard wages. Any surplus after meeting: (a) Standard Wages; (b) Costs of Production other than wages; and (c) Standard Profits; was to be divided between wages and profits in the proportions eighty-three to seventeen.

The percentage additions to basis wages in the Eastern District varied widely according to its state of trade as the following table shows:

PERCENTAGE ADDITIONS TO 1911 BASIS RATES PAID DURING LIFETIME OF EASTERN DISTRICT

Month	1921	1922	1923	1924	1925	1926
January		109·86	49·61	55·66	60·76	46·67
February		97·06	49·61	55·66	58·66	46·67
March		91·23	45·21	58·47	57·47	46·67
April		80·37	45·21	58·47	57·58	46·67
May		79·71	54·52	77·44	57·04	
June		64·43	54·52	77·44	50·62	
July		53·78	58·56	79·83	46·67	
August		32·00	58·56	68·37	46·67	
September	131·00	36·19	58·37	60·28	46·67	
October	140·19	39·04	58·37	52·53	46·67	
November	110·55	43·31	47·63	57·18	56·67	
December	109·54	46·69	47·63	59·75	46·67	

Source: Wages, 1950, issued by NUM (Notts. Area), p. 3.

According to the Nottinghamshire Guardian of July 4, 1921, Mr F. B. Varley had 'warned the Nottinghamshire miners' at the outset of the 1921 dispute 'that they could not carry the fight through to a successful issue', and for this reason, alone of the members of the MFGB Executive Committee he had 'publicly urged a settlement'. There is no doubt that the very heavy reduction in wages which the deepening depression of 1922 brought, indicating as it did the complete failure of the 1921 Lockout, encouraged the growth of apathy and even of positive anti-unionism amongst the Nottinghamshire miners. Further, those of them who understood that things might not have been so bad had the County formed a separate wages district may be forgiven for questioning the official Federation policy of a national pool which would have made things worse still.

H

CHAPTER 6

DEPRESSION AND RECOVERY

1. *Depression*

The deepening of the depression in 1922 hit the miners hard. While it is true that prices were falling, the miner's wage-level fell much further and much faster. The higher rate of unemployment hit miners in common with other workers, but the fall in wage-rates was far more marked in mining than in most other occupations; so that the miner's position relative to that of workmen in other industries was seriously worsened. This is borne out by the table below:

CHANGES IN WEEKLY WAGE-RATES

Trade	Aug. 4, 1914	Apr. 30, 1919	Feb. 29, 1920	Feb. 28, 1922	Sep. 30, 1922	Mar. 31, 1923	Cost-of Living at Mar. 1923 (1914 = 100)
BUILDING:							
Bricklayers	100	188	223	214	176	176	
Labourers	100	299	284	254	198	198	
SHIP-BUILDING							
Riveters	100	198	213	189	146	120	
Joiners	100	254	278	248	175	169	
ENGINEERING:							
Fitters and Turners	100	198	212	189	148	145	
Labourers	100	256	280	250	177	177	
IRON AND STEEL							
LABOURERS:	100	250	250	220	200	200	174
RAILWAYS:							
Porters	100	283	310	300	260	260	
Foremen	100	183	200	207	185	185	
COAL MINERS:	100	215	223	150	130	130	
WOOL & WORSTED:							
Spinners and Weavers	100	207	225	214	185	185	
COTTON:							
Spinners and Weavers	100	205	205	205	169	161	

Source: *Manchester Guardian Commercial*, 'European Reconstruction', Section 16, p. 867; quoted Pigou, op.cit., p. 233. Notice how the lower-paid workers improved their relative positions during the war.

Since industrial action did not seem, for the moment, to be productive of much good, the NMA took a more active interest in politics. We have already noticed that the MFGB had within its ranks on the one hand men who believed primarily in the efficacy of direct

industrial action; and on the other, people who believed that fundamental changes could only be achieved through the ballot box.

This division ran right through the Labour and Trade Union
Movement in the early 1920s. As we saw in Chapter 2 there was a
pronounced swing to the Left during and after the war. The Russian
revolutions of 1917 contributed materially to the climate of opinion.
The knowledge that the tyranny of the Czars had been brought to an
end by the working class acted as a ferment among the trade unionists
of the West.

However, the formation of Communist Parties in other countries
made inevitable a deep and serious split in the Labour Movement.
The Left-wing, Communist-led, sought opportunities for sharpening
the class conflict, believing that capitalism had been so badly shaken
by the War and by the post-war dislocation of markets that it could
not last much longer in the face of determined opposition by the
organized working class. The moderate Right-wing of the Labour
Movement on the other hand, saw the period as one in which the
worker was bearing far more than his fair share of the burden imposed
by the universal economic malaise; they therefore sought to effect a
change of Government by constitutional process whilst, at the same
time, missing no opportunity to secure improvements of the worker's
conditions.

In 1922 the South Wales Miners' Federation tried to bring the
MFGB within the orbit of the Communist Movement. The following
motion appeared on the agenda of the Annual Conference on
Wednesday, July 18th:

> 'That the Miners' Federation of Great Britain be urged to
> affiliate and actively identify itself with the Third International.'

Mr S. O. Davies, the mover of the motion, explained that 'Third
International' should read 'Red International of Labour Unions'.
The seconder of the motion, Mr A. J. Cook, explained the difference
between the 'Red' International and the old-style 'Yellow' International (the International Federation of Trade Unions) thus:

> 'The difference in principle is that the Red International desires
> to abolish capitalism, but the other simply appeals to it. There
> are many leaders prepared to do that, but what is oratory without
> the necessary force behind it.'

Later: 'I do not want to fight with guns, bladders are bad enough for
me. I believe in the organized might of the working classes. I believe
in the workers' might, therefore do not be afraid of a revolution. With
regard to the Red Trades Union formed in Russia, they have a
membership of several millions now, formed with the definite object

of preparing for the international overthrow of capital, and if that fails to become operative then we must be prepared to face losses in wages.'

The motion was lost, only South Wales voting in favour.

Nottinghamshire was thought of as being a 'moderate' county but even so, as we have seen, it had its quota of militants. People like George Spencer and F. B. Varley had themselves been regarded in their youth as Left-wing, and still, on occasion, made speeches with a Left-wing flavour. At a meeting held in Hucknall in connection with the return to work following the 1921 Lockout for example, Spencer spoke of the men being driven back by hunger; and he denounced both the coalowners and the Government in a downright manner. The remedy which he advocated however, was mild: people must be taught to vote Labour at future General Elections. To the right of Spencer and Varley stood J. G. Hancock who sat in Parliament as Liberal member for what was now known as the Belper division until 1922. On November 29, 1924, the Summit delegate moved a vote of censure on Hancock for 'his action in supporting the Tory candidate at the recent election'. Summit went on to ask that Hancock, together with Charlie Bunfield and William Carter, should be asked to resign their positions. Neither of these motions was adopted. Motions of this kind display the continuing influence of Jack Lavin's IWW propaganda.

The Mansfield District Committee (or Mansfield Area Committee) which came into existence in this period was the organizing centre of the Left-wing. Many of its members were infected, to use Mr Herbert Booth's term, by the 'Lavin virus', whilst others were Left-wing socialists of a more orthodox cast. This committee grew out of a committee set up in connection with the Berry Hill Hall Convalescent Homes for Miners. The convalescent home was financed by the Nottinghamshire and Derbyshire Miners' Associations; and the committee consisted of two representatives from each branch with members living in the Mansfield area. By the end of 1922, Berry Hill had become a 'White Elephant', the miners preferring to go to the seaside, but the committee continued to meet to discuss political and industrial matters at the Mansfield Labour Club. Its political activities apart, this organization continued the fight against the 'butty' system.

However, the butties had an organization of their own, and this was also centred on Mansfield. The leader of this organization was William Holland, a former Vice-President of the NMA, who addressed meetings all over the county and in the Bolsover-Creswell area of Derbyshire. Associated with William Holland were Joseph Birkin, S. F. Middup, H. Willett, and later, Wilfred Stevenson (who later worked for the Economic League) and S. Thompson. This organization had the backing of some colliery owners and in the period of depression following the 1921 lockout it met with a large measure of success in re-establishing the butty system—first in the Mansfield district and later in other parts of the county. This body was linked up with a would-be national movement, the 'British Workers' League',[1] which advocated non-political trade unionism and peaceful relations with the owners. The Economic League also moved into the coalfield and the two organizations so overlapped that it is difficult to distinguish between them. The Economic League continued its work, on behalf of the owners, up to the time of nationalization.

By 1926, the butties' organizations covering such important pits as the Bolsover and Barber Walker groups, were firmly established and they were to give George Spencer the nucleus of his Industrial Union besides supplying him with a ready-made programme (separation from the MFGB; non-political unionism; freedom from strikes; and the consolidation of the butty system).

Meantime, whilst Spencer was still a loyal member of the Parliamentary Labour Party, Joseph Birkin, William Holland and company were already at work with their anti-political union propaganda and, in November, 1922, Council resolved:

'That the Financial Secretary obtain from the records the amounts drawn from the Political and General Funds by Mr J. Birkin and William Holland.'

A month later, Council further resolved:

'That the Financial Secretary get all information with regard to statements made by persons in the County concerning the administration, finances and any other side of our activities with a view to taking legal proceedings if necessary.'

One particular criticism raised by Right-wing members of the union concerned the heavy cost entailed in sending students to the Central Labour College: the syndicalist seminary from which has grown the National Council of Labour Colleges.

In our last two chapters, we saw how membership fell following the collapse of the 1921 Lockout. This fall in membership was disconcerting, especially to the 'direct actionists', and a strenuous campaign to bring the men back into the union was undertaken. At first, the campaign was sedate: meetings were addressed by the district officials of the Association. Later, the 'Back to the Union' Rallies assumed the appearance of revivalist meetings with A. J. Cook as the evangelist in chief. The men were now asked to join the union on a Left-wing programme. This campaign deepened the division in Nottinghamshire since Cook, after his election to the General Secretaryship of the MFGB, was able to address the rank-and-file of the NMA over the heads of their district officials who were often opposed to him on questions of policy. We shall examine the situation in more detail in our next section, but enough has been said to demonstrate the unwisdom of Cook's campaign, to which some of the blame for the 1926 'split' must be attached.

2. Recovery

On Tuesday, February 27, 1923, the MFGB Executive Committee had before it an appeal for assistance for the workpeople of the Ruhr who were suffering hardship as a result of the French occupation of the district in which they lived. The Federation had already made a grant to the Ruhr miners; and the Committee decided that no further help could be given.

However, the French occupation of the Ruhr certainly brought much-needed relief to the British Coal Industry; and what the German miner lost on the swings his English counterpart gained on the roundabouts. Mr Page Arnot says: 'The year 1923 which found most industries in a state of depression following on the 1921 slump, actually found a short-lived boom in coal exports.' Actually, December 1922 saw the end of the slump and the beginning of what Professor Pigou calls 'the doldrums'. The percentage of unemployment revealed by the figures compiled from Trade Union returns fell from 13·6 in January 1923 to 7 in May 1924. The demand for coal for both inland consumption and for export therefore rose as the following figures indicate:

Year	Output of Coal	Total Quantity of Coal Shipped Abroad	Coal Available for Home Consunption
1922	249,606,864	87,351,530	162,362,662
1923	276,000,560	102,817,570	173,303,021
1924	267,118,167	81,742,314	185,485,601

Source: Colliery Year Book, 1951 Edn, p. 576.

Output in the counties of Nottingham, Derby and Leicester combined similarly rose from 30,772,057 tons in 1922 to 34,916,672 in 1923, and fell slightly to 34,189,686 in 1924.

The improvement in trading conditions which manifested itself in 1923 was used by the MFGB to try to secure an improved wages agreement. George Spencer and Frank Varley opposed the majority view however. At a Special Conference of the Federation held on Tuesday, March 27, 1923, Lancashire and South Wales proposed: 'That this Conference . . . recommends that districts be at once con-

sulted upon the termination of the existing National Wage Agreement . . .' To this Varley moved an amendment: '. . . that districts
take whatever steps they deem effective to obtain the opinions of
their men as to the continuance or termination of the agreement, and
report to a further Conference . . .' In moving this amendment he
made it quite clear that he was against the termination of the agreement. Lancashire had argued that under the existing agreement, a
street sweeper was getting more money than a skilled miner; this
argument Varley dismissed as irrelevant: '. . . a street sweeper of
Bury has more than the miners of Bury at this particular time because
the Corporation of Bury have a monopoly of the streets.' He thought
that it would be 'a gross betrayal' to commit the men to 'precipitate
action'. Varley went on:

> 'I do want to appreciate the difficulties which lie ahead, but it
> must be remembered when four weeks had expired in the last
> national stoppage, whilst we continued the struggle to raise the
> minimum wage and increase wages, some believed that it would
> not possibly bring one single penny more no matter how long we
> stood than if we had gone back at the four weeks' end. We kept
> the men out, and in the process of so doing we built up a debt of
> £163,000 for less than 43,000 men. We still owe £80,000.'

After referring to the decline in membership, he went on:

> '. . . I am as confident as I stand here, having regard to the fact
> that the last stoppage so strained the loyalty of our men, *if we
> attempt to strain that loyalty again, it will smash us.*'

Mr Hough of Yorkshire pointed out that:

> 'Neither at Southport in December, nor at London in March
> has anything been suggested to substitute (for the Agreement).'

This was, indeed, the kernel of the case advanced by the 'moderates': it is all very well to talk about scrapping the Agreement, but
can we really expect to get anything better? Further, are we strong
enough to hold out for anything better? At a further conference,
held on July 10, 1923 at Folkestone, Varley enlarged on this theme
thus:

> '. . . You know perfectly well, whilst our machine was strong we
> did not succeed; and shall we succeed whilst our machine is in its
> present state? I believe we should smash the machine we have got.
> I am not in such a pitiable condition as Mr Latham (Midlands)
> finds himself, nor Lancashire, but I have sympathy with them, and
> to show that we have sympathy with them we desire to continue

the present agreement. If it were smashed, it would show we had no sympathy with them. Were it not for the operation of this agreement, Shropshire would have found itself outside the competitive market.'

Varley then referred to the Lancashire suggestion that 'For the purpose of determining the profits of the employer, regard shall be had to the aggregate capital invested, and the employer shall be entitled to a fair return thereon. But such profits shall not be calculated in relation to the standard wages of the workmen'. He compared the capital invested in the Kent coalfield (about £3½ million) with that invested in Mansfield Colliery (about £400,000) and asked whether the owners in these two cases would receive profits related to the size of their respective investments, bearing in mind the respective outputs: less than 500,000 tons a year for Kent compared with Mansfield Colliery's 1913 output of 1,100,000 tons.

He then returned to the consideration of strategy:

> 'We, the Nottinghamshire Miners' Association, owe something like £65,000, and therefore, if the principles of this agreement were foisted upon us as he (Mr Latham) says, at a time when by comparison, we were stronger than we are now, what can we hope to obtain if we cease work in the same way?'

This pacifism roused the ire of a Scottish delegate, Mr McNulty, who said: 'I want to refer to the subtle way things have been put. I have found in my experience during this wages' campaign that the men who are opposing this motion today are the same men who oppose every aggressive movement that emanates from the rank-and-file. . . . The miner does not come on strike with a chest full. . . . There is no stated time when we should strike; we do not wait while we have money in the chest. If we intend to fight, it should be now, and despite the doleful tales we have heard, we ought to arrange to fight this year.'

Mr Peter Lee of Durham said frankly, speaking of the original programme for a national pool:

> 'We also went in for another thing, we went in for a pool, whereby the better-paid districts would help the poorer-paid districts, and I venture to suggest that if the pool had been applied under the present agreement so far as we are concerned in the Miners' Federation there would not have been so much dissension as at the present time.'

Needless to say, Durham would have been a beneficiary from any pooling scheme arrangement.

Following this conference, the MFGB Executive entered into negotiations with the owners with a view to securing an improved agreement rather than an agreement with fresh principles.

In November 1923, the NMA Council mandated its delegates to a national conference to be held in the month following to insist that before any decision were taken on the question of terminating the Agreement, the matter should be referred to districts for consideration; and that a ballot vote of the members should be held. However, if after all, the MFGB Conference were to take a decision without referring to districts, then the Nottinghamshire delegates should vote in favour of the continuation of the old agreement. This decision was no doubt due to the very strong views held by Varley and Spencer and which were doubtless shared by their Nottinghamshire colleagues.

However, the decision taken by the MFGB Conference was: 'That a ballot vote be taken and that the men be advised to vote in favour of terminating the agreement.'[2]

When the results of the ballot vote became known, it was found that the members of the NMA had voted in support of the National Conference's recommendation to terminate the Agreement, despite the stand taken by their district officials. The voting was as follows:

NATIONAL:	For terminating the agreement	510,303
	Against terminating the agreement	114,558
NOTTINGHAM:	For terminating the agreement	11,392
	Against terminating the agreement	5,059

However, it should again be borne in mind that the 16,000-odd people who voted in Nottinghamshire formed little more than one-third of the total labour force.

In the early part of 1924 Varley and Spencer came under attack for making public statements on wages policy which ran contrary to the National decision.

At its meeting on December 29, 1923, the NMA Council had resolved: 'That we adhere to the Federation Conference decision and recommend members to vote in favour of terminating the present wages agreement', thus reversing the decision taken on November 24th. Further, Council confirmed its support for a new national agreement at its meeting on January 26, 1924. Between these two meetings however, Varley and Spencer had continued to express their conviction that the existing agreement should continue in force; with modifications probably, but with the fundamental principles unchanged. The Left-wing members of Council criticized these public statements, but no further action was taken.[3]

The MFGB gave three months' notice to terminate the 1921 Agreement on January 17, 1924. Negotiations with the owners ambled on until March 12th, when the Mining Association offered the following terms:

1. Standard profits shall consist of 15 per cent of standard wages instead of 17 per cent.

2. The surplus left after meeting standard wages, standard profits and costs other than wages shall be divided in the proportions 87 per cent to wages and 13 per cent to profits instead of 83 per cent to wages and 17 per cent to profits.

3. The general minimum percentage on standard wages shall be 30 per cent instead of 20 per cent.

During the debate on this offer, which took place at a special Federation Conference on Thursday, March 17th, Mr Varley was once again criticized for his public pronouncements. Mr G. Griffiths of Yorkshire alleged that Varley had said, at a public meeting held over the previous week-end, that if the owners offered a shilling a day increase we should accept it. Mr Griffiths implied that this statement ran counter to official Federation policy, and as a member of the Negotiating Committee, Mr Varley had no right to make statements of that kind. Varley explained that:

'I was speaking in Mansfield, and I was saying that the dockers had won a remarkable victory, and that there was no reason why the miners' efforts should not be attended with similar results. One of the persons said they had not won a victory, but that the [Labour] Government had let them down. Instantly I said that if His Majesty's Government would let us down to the extent of giving us one shilling now and one shilling in July I would come back and advise acceptance.'

Actually, under the terms submitted by the owners, Nottinghamshire (on Varley's calculations) stood to gain only 3¾d a shift, whilst those districts to which the minimum wage conditions would apply: Scotland, North Wales, and so on, would get far more. At the adjourned Federation Conference held on March 26th and the succeeding days Varley moved, on behalf of the Executive Committee:

'That it be a recommendation to reject the terms now offered as providing no solution of the miners' wage question generally, nor giving the immediate relief so vitally necessary, and calls upon the Government to institute an inquiry into:
'(a) The wages of miners compared with their 1914 wages;
'(b) compared with the Cost-of-Living index figure; and
'(c) compared with the wages now prevailing in industry generally.'

In moving this motion for the rejection of the owners' offer Varley said '. . . that it might have been thought I should be supporting a recommendation for acceptance. That would have been perfectly pardonable inasmuch as of the whole of the members of the Committee, I have been deemed most pacifist.' He also made his motive in opposing acceptance of the offer clear when he declared: 'I do not want to minimize in the slightest degree the benefit which the application of these proposals would give. I would go so far as to say this, that if I were here representing any other district than my own, I should have very serious hesitation in doing what I am, because no man can afford to lightly disregard an increase of 10d a day to his men.'

Although Varley's motive in supporting the motion of the Executive Committee was, therefore, seen to be selfish he went on to say that the Executive Committee did not suggest strike action as an alternative to acceptance of the offer; but instead, thought that an inquiry would be helpful since the miners had a strong case, and since a Labour Government was in office.

The motion was supported by Greenall of Lancashire and A. J. Cook of South Wales, both of whom were usually in the opposite camp to Varley. Mr Robson of Durham, who was in favour of accepting the owners' terms, thought that Cook was largely responsible for the weak position in which the miners found themselves since he and his kind had rushed the Federation into a dispute in 1921 which had sapped its strength; whilst Thomas Richards of South Wales, acting as Chairman of the conference in the absence of Herbert Smith, thought Cook and Varley strange bedfellows:

'. . . from the first minute we entered into the negotiations I have had all my work cut out to keep peace between Varley and Cook. They were opposite in their conception of how the problem should be treated. Now some miracle has taken place—they are both supporting the same resolution.'

One finds it strange that Richards could have spoken against the recommendation of the Executive Committee without being forced to vacate the chair, but he did so.

In his reply to the debate, Varley said that he was never more than 'a half-way house man' who deplored strikes. Nottinghamshire and the Eastern Division generally had a right, however, to benefit by any new agreement. Nottinghamshire had suffered for the lower paid men in 1921—having come out of the Lockout £173,000 in debt, of which £47,000 was still owing—and now he wanted 'something to take back to Nottinghamshire'. He was convinced that 'if we can get this inquiry we have hope in the present psychology of the British public, of accomplishing something better than that (i.e. the owners' offer). I do want to be fair and straight, I believe we can do better, we cannot do worse and it was for that I, in all sincerity, moved that resolution.'

The conference accepted the Executive Committee's recommendation; and subsequently the men, in a ballot vote, did so too. The voting was however, very close. Nationally, 322,392 voted in favour of accepting the owners' proposals and 338,650 voted against. The figures for Nottinghamshire were: 7,909 for acceptance of the offer and 7,506 against. The Left-wing districts, South Wales and Lancashire on the other hand, recorded huge majorities against acceptance of the offer, despite the fact that they stood to gain by it.

The result of the ballot was conveyed to Mr Shinwell, Secretary of the Mines Department, who agreed to establish a Court of Inquiry under the Industrial Court Act of 1919. The members of the Court were:

Lord Buckmaster, Chairman;
Roscoe Brunner, JP (an industrialist); and
H. G. Cameron (Secretary of the Amalgamated Society of Wood-
    workers).

The Court sat for five days and presented its report (*Command Paper* 2129) on May 8th. As Arnot says, the wording of the report indicates that the MFGB had substantially proved their case for improved wages and 'concluded with the recommendation that negotiations should continue with a view to a modification of the terms of the agreement of 1921'.

At a special conference of the MFGB held on May 29, 1924, the new Secretary of the Federation, Mr A. J. Cook, reported on the subsequent negotiations. The Federation had really made very little advance on the owners' previous offer. The owners had now agreed that standard profits should be 15 per cent of standard wages (as previous offer); the surplus should be divided between wages and profits in the proportions 88:12 (previous offer 87:13); and the general minimum should be $33\frac{1}{3}$ per cent on standard wages (previous offer 30 per cent). Conference agreed to accept the terms and that the revised agreement should run for twelve months.

Some difficulty arose in applying this Agreement in the small coalfields of Kent, Bristol and Forest of Dean, where conditions were such that the owners could ill afford to concede anything. Mr Straker and Mr Varley put forward the classic argument of the MFGB on this question—if a pit cannot afford to pay a decent wage it should go out of production. As Varley put it:

'We had it in our minds under private enterprise that those pits which could not pay the wages dictated by the National Agreement would go out of production. Therefore, if the Conference will say quite frankly what they really believe, and be frank and recognize this, that if pits cannot pay they must go out of production, then I think we shall have to employ the last alternative.' [i.e. to assist the men financially to stay unemployed until an improvement in trade once more took their districts above the margin of production.]

On much the same grounds, Spencer, speaking at the Annual Conference of the MFGB on Friday, July 11th, 1924, opposed a motion:

'That the Miners' Federation of Great Britain endeavour to secure an agreement for the establishment of a National Wages Board for the regulation of wages over the whole of Great Britain, and put all the members of the Federation on one basis, instead of the present district ascertainment.'

Spencer said that if they were to be successful in putting 'all the members of the Federation on one basis' then the less profitable districts would be forced out of production:

'It must be inevitable if we have an ascertainment which is giving an equal wage for each one of us, that is on a uniform average, then the least economic districts will go out of existence. If Lancashire, Bristol and Somerset have to pay a wage equal to the Eastern Area wage they cannot possibly carry on, and therefore I consider it is futile and oppose it.'

He considered the question of the national pool to be a political one, which would have to be dealt with, if at all, by political action and not as in 1921, by industrial action. If a national pool were to be introduced, then a national ascertainment of wages would be practicable. One point which Spencer did not make on this question, but which he made on many subsequent occasions, was that a national ascertainment would involve the subsidization of the poorer districts by the wealthier ones.

The Right-wing stand taken by Spencer and Varley in this period seems to have made them unpopular at national conferences. This, at least, is how one reads the results of the voting for various offices. Thus, at the Conference held at Swansea on July 8th and the succeeding days, they came at the bottom of the poll for the posts for which they were candidates. The details are:

TUC GENERAL COUNCIL (2 REQUIRED)

Candidate	District	Votes Polled	Voting Strength of 'Home' District
R. Smillie	(Scotland)	597	80
T. Richards	(S. Wales)	479	148
P. Lee	(Durham)	204	120
E. Hough	(Yorks)	148	140
F. B. Varley	(Notts)	125	25

LABOUR PARTY E.C. (2 VOTES)

Candidate	District	Votes Polled	Voting Strength of 'Home' District
Arthur Jenkins	(S. Wales)	523	148
J. Jones	(Yorks)	475	140
J. Herriotts	(Durham)	247	120
H. Hicken	(Derbys)	229	32
G. A. Spencer	(Notts)	38	25

Source: MFGB *Minute Book,* 1924, p. 562. I accept, of course, that
the size of a candidate's home district, which will inevitably vote
for him, counts a lot; but it cannot account for Spencer's
insignificant vote.

In much the same way, their popularity waned in their own district,
even though they had both been successful in the parliamentary
elections of December 1923. Varley, who was MP for Mansfield,
always found his meetings lively affairs; and indeed Mansfield, the
centre of the new coalfield of large pits working the 'Five Foot Top
Hard' Seam, was the centre also of Left-wing activity co-ordinated
by the Mansfield District Committee.

The Mansfield District Committee organized rallies in connection
with A. J. Cook's 'Back to the Union Campaign'. Summit Lodge,
which had as its delegate William Bayliss, later to become a full-time
official of the union and Chairman of the Nottinghamshire County
Council, requested permission to hold two recruiting meetings to be
addressed by Cook as early as May 31, 1924, but this request was
turned down. However, at the August Council Meeting, Lowmoor
branch wanted to know who had borne the cost of the Mansfield
'Get Back to the Union Meeting', and what payment the individual
speakers had received. (The Lowmoor delegate at this time, Mr R.
Gascoyne, acted as secretary of the Spencer Union upon its formation in 1926.) At the following Council meeting, Mansfield branch
put forward an appeal: 'I am requested by mandate from our general
meeting to appeal to Council for their sanction for the Area Committee (Mansfield and District) to be allowed to continue their

'Back to the Union' campaign'. The demonstration at Mansfield, against which Lowmoor had appealed, had cost a total of £37 2s 1d (including a payment of £5 to A. J. Cook) which was shared jointly by Nottinghamshire and Derbyshire.

The Mansfield campaign continued, becoming a battleground between Left and Right. Thus, from Rufford (Left-wing) came a request that Herbert Smith should address a 'Back to the Union' meeting, whilst two months later Lowmoor (Right-wing) requested that Bob Smillie should similarly be invited. Again in June, Rufford asked that A. J. Cook should be engaged to address mass meetings in defence of the eight-hour day, whilst a month later, Bestwood similarly requested the services of Frank Hodges who was now travelling further and further to the Right.

Rufford, Summit and Welbeck kept up a barrage of Left-wing appeals to Council; and indeed, most of Council's time appears to have been taken up by political matters in this period. The following motions are samples of the Mansfield district's output:

January 31, 1925: Rufford appeal for (*a*) permission for branches to affiliate to Mansfield and District Labour College and the Labour Research Dept, and (*b*) for May 1st to be made into a National Holiday to 'bring us into line with the International Working Class programme'.

Welbeck congratulate A. J. Cook on his 'courageous Back to the Union' campaign. Welbeck put forward a suggestion for One Union in the Nottinghamshire coalfield.

March 28, 1925: Rufford ask for financial support for the *Daily Herald*'s circulation campaign.

April 25, 1925: Rufford ask that the General Council of the TUC should 'take immediate steps to co-ordinate the Forces of the Allied Trade Unions (Miners, Engineers, Railwaymen and Transport Workers, etc.) to joint action to enforce the rights and demands of the *workers generally*'.

June 27, 1925: Rufford urge the TUC to take steps to prevent the Government from armed intervention in China.

Undoubtedly, the tide was now running strongly leftward although the district officials did their best to stem it. The fiery speeches of A. J. Cook, expressing as they did the revolt of the miner against a harsh environment, captured the imagination of the mining population. Cook's appeal however, was an appeal to the emotions: his own as well as his auditors'. Many of Cook's prepared speeches at MFGB Conferences were comparatively mild in tone compared with his speeches in the country. This displeased the Left, so that we have Mr McNulty of Scotland (a Communist) saying:

'Now, the miners in the country are looking for a lead from us here. Mr Cook has been the most outspoken individual in connection with this movement, but it is no use saying things outside unless they are going to be said here. . . . I am not casting any reflection on what Mr Cook has said, but he has said things outside which he has not said here.'

At the same conference, Mr Greenall of Lancashire made the same complaint, and when Mr Cook denied that he had said what had been attributed to him, Greenall offered to send him a copy of his speech from the report of the meeting.

Spencer was also displeased, but from the opposite angle. 'I would like to ask the platform,' he said, 'whether they endorse what Mr Cook has been saying Sunday after Sunday with regard to a revolution. The language which has been used is going to break this Federation. I shall myself protest about it outside. It does not make for unity.'

> MR COOK: 'I am coming to Mansfield, you can do it there.'
> MR SPENCER, MP: 'Is the platform in favour of the statements which have been made?'

The Federation was about to embark on its greatest struggle. Well might New Hucknall branch appeal: 'That we deplore the methods of our leaders wrangling in public and the press, as the same gives the public and the press the opportunity of revelling in such matters when taken up by the press.'[5]

If 'Unity is Strength', then division is weakness. In 1925 this simple corollary appears not to have been heeded.

APPENDIX
(See Minute 5)
NOTTINGHAMSHIRE AND DERBYSHIRE MINERS' DEMONSTRATION
held July 19, 1924

Expenses		£	s	d
Printing per Willman	50 D.D. Posters	1	4	0
	50 Small Posters		11	6
	15,000 Handbills	4	10	0
	50 Circular letters		6	6
	100 4 pp Programmes	1	4	0
	750 Handbills		8	6
Plumb and Richardson—Badges			8	0
Meetings and Chairs, Labour Club, Mansfield			16	6
Meeting, 'Bowl in Hand', per H. Oliver			7	6
Sherwood Bill Posting Company			13	9
Refreshments for Bandsmen		1	9	0
Retaining Fee, Hippodrome		1	0	0
Mansfield Excelsior Band		5	0	0
Shirebrook Silver Band		5	0	0
Shirebrook Silver Band Railway Expenses		5	0	0
Stamps, Leaflet Postages, Calling Meetings, &c.		1	0	6
Haulage for Platform of Chairs and Forms			7	6
Secretary's Petty Cash			3	0
Secretary's Lost Time		2	10	0
Langwith Banner Carriers		2	0	0
Secretary's Remuneration		2	0	0
Total:		£32	2	1

To Nottinghamshire Miners' Association,

We, the undersigned, hereby certify that the above Statement is correct, and bills for same have been presented accordingly.

JNO. CLIFF, Welbeck.
S. GARNER, Clipstone.
WM. SMITH, Crown Farm.
G. ABBOTT, Secretary.

	£	s	d
Brought forward	32	2	1
Paid to A. J. Cook	5	0	0
Total:	£37	2	1
Derbyshire Share	18	11	0
Nottinghamshire Share	18	11	1

CHAPTER 7

PRELUDE TO DISASTER

1. *The Economic Background*

The 'recovery' of which we wrote in Chapter 6 was partial and temporary. Even at its peak in May 1924, the unemployment rate was still 7 per cent. By the end of 1924, 9·2 per cent of trade unionists were out of work; and by the end of 1925 the figure was 11 per cent. Some industries were hit worse than others, heavy industry being much the most badly affected, as the following examples demonstrate:

UNEMPLOYMENT, 1924 & 1925 (HEAVY INDUSTRIES)

Industry	Number Employed	Percentage Unemployed	
	1923	1924	1925
Coal Mining	1,244,000	5·7	15·8
Ship-building & Repairing	270,000	29·4	33·8
Steel Smelting, Iron & Steel Mills, etc.	211,000	21·1	24·5

Source: Pigou, *Aspects of British Economic History* 1918–25, p. 50.

Look for comparison at the following representative consumer goods industries:

Industry	Number Employed	Percentage Unemployed	
	1923	1924	1925
Furniture Making, etc.	94,000	7·2	6·2
Motor Vehicle, etc., Manufacture and Repair	192,000	8·5	7·0
Boots, Shoes, etc.	142,000	9·2	10·5
Bread, Biscuits, etc.	160,000	9·4	9·1
All Industries	11,486,000	10·3	11·3

Between May and November 1924, some 111 collieries had to close down. The small Kent coalfield, in February 1925, had one pit working with a labour force of 450 instead of the usual 1,000; whilst another company with a nominal capital of £750,000 had the Official Receiver in and its £1 shares were changing hands at 1½d each. The funds of the NMA were again having to bear unemployment benefit of £300 to £400 a month; this despite the Union's continued indebtedness.

The coalmining industry was suffering from the general decline in World trade; and in addition, alternative sources of power: hydro-electric power, oil and lignite were now being developed. Further, new fuel consuming plant tended to be more efficient than the old, thus needing a smaller input of coal for a given output of energy. The Miners' Federation was also concerned at the effect of reparation coal (i.e. coal supplied by Germany as reparation for war damage) upon Britain's traditional European markets; whilst the Mining Association alleged that Britain's share of total World trade in coal was declining. This last supposition cannot, however, be reconciled with the facts. As Prof. Tawney has pointed out, 'the British share of the total world export was slightly higher in 1924 than in 1913— 51·2 per cent as against 49·8 per cent'. However, in absolute terms, Britain's coal exports 'fell from 88·37 million tons in the years 1909– 13 to 68·97 million tons in the year 1925, and foreign bunkers from 19·6 million tons at the first date to 16·2 million tons at the second'.[1]

Government policy in this period has been described by Prof. Pigou thus:

'. . . the state of the American exchange, rather than the industrial situation at home, was their principal pre-occupation. A necessary preliminary to the restoration of the Gold Standard was to bring the exchange back to the neighbourhood of par. When it began to run away, measures had to be taken to arrest it; when it tended to stand still, further measures to improve it. These measures took the form of operations on the discount rate and Bank dealings in securities aimed at reducing the English, relatively to the American, price level. When that price level was itself falling, this entailed reducing our own price level in an absolute sense and in a greater degree. Any restrictive action which the Americans took we must take also; and, even when they did not take any, we, to make up our leeway had, nevertheless to take some.'

During the winter of 1924–25, the exchange improved, helped by the operations of speculators who knew that the restoration of the Gold Standard was in sight. At the end of February 1925 the New York re-discount rate was raised from 3 to 3½ per cent, and the Bank of England followed suit by raising the Bank rate to 5 per cent. Finally, on April 27th the Chancellor of the Exchequer announced that 'the Bank of England would sell gold for export without restriction at the coinage price of £3 17s 10½d a standard ounce' (that is, its pre-war price). Britain had returned to the Gold Standard; sterling was back on its throne as the international currency par excellence; the national prestige was enhanced. But at what cost.

Writing in July 1925 Sir Josiah Stamp, member of the Macmillan Court of Inquiry concerning the Coal Mining Industry[2] declared:

'In my own judgment, with all due appreciation of the virtues of the Gold Standard as such, it was always open to doubt in the period January to March 1925 whether its introduction would not involve a considerable degree of actual deflation . . . and whether such deflation would not inevitably be so one-sided in becoming effective as to cause much industrial unrest.'

Sir Josiah points out that the position of sterling relative to the dollar was artificially high owing to the actions of speculators as well as to 'special causes relating to money rates in London designed to retain foreign balances here', and he then proceeds to explain the special difficulties created for the coal trade:

'The Times Index of Prices shows a continuous drop for six months—January to June—from 179 to 155—a fall of nearly 14 per cent. It is quite obvious that if the British exporter has to face a diminution in his receipts per unit of production upon this scale, and has no power to secure an immediate reduction on his outgoings (other than raw materials) he must work at a severe loss, either by actually selling goods at a loss or by quoting too high prices and not selling goods at all. . . . It ought to be obvious that those of our exports on which there are fine margins in close competition would be the first to feel the pinch. The coal industry, starting from the point of having little margin of profit at all into which to cut, would readily be put into serious difficulties. Moreover, coal is sensitive to foreign competition to a peculiar degree, because it not only suffers its own direct troubles in a curtailment of exported coal, but it feels the reduction in other classes of exports, such as iron and steel, in the most immediate and direct way. In my view therefore, the recent improvement of the exchange or decline in the price level to which I have referred, whether or not compulsorily brought about by the anticipation and then the realization of the gold standard is sufficient in itself to account for the *special* plight of the industry since March.'

Keynes, writing some little time after Sir Josiah Stamp, took much the same line. He pointed out that Britain's depression was unique: other countries (after having realistically devalued their currencies) were enjoying a period of prosperity and he declared that our troubles were due to the high prices of our exports relative to the prices charged for similar goods by our competitors. He refuted the view that our high prices resulted from the allegedly high wages and low productivity of British workmen. Instead, '. . . the explanation can be found for certain in another direction. For we know as a fact that the value of sterling money abroad has been raised by 10 per cent, whilst its purchasing power over British labour is unchanged. This alteration in the external value of sterling money has been the deliberate act of the Government and the Chancellor of the Exchequer, and the present troubles of our export industries are the inevitable and predictable consequences of it'.[3] Keynes warned that the Government's policy involved the cutting down of wages and other expenses by 10 per cent. Eventually price levels in Britain would fall, but in the meantime those workers whose wages were to be cut immediately would suffer a sharp drop in their standard of living.

The return to the Gold Standard had had a deplorable effect on our coal exports. We had been priced out of the South American, Canadian and European markets and in order to bring our prices back on to a competitive level it would be necessary to reduce them by 1s 9d a ton. This could not be met from profits, since these amounted to only 6d a ton in the first quarter of 1925. The gap could only be bridged by reducing miners wages. Thus the miners were to be made to suffer the consequences of a policy for which they were in no way responsible.

It would be foolish to argue that the coal industry's difficulties in this period were due solely to the effects of the return to the Gold Standard at the pre-war parity. The industry was suffering from the competition of alternative sources of power (lignite, oil and hydro-electricity); and from wasteful competition in the fields of production and distribution. There were too many inefficient pits in production and too many miners seeking work. Further, our pits were slightly less efficient than those of Upper Silesia and Poland, and much less efficient than those of the USA. At the same time, because the industry was already making heavy weather, the improvement of the pound could be expected to hit the mining industry far more severely than other industries where these special factors did not operate.

The chief complaint of the owners in this period was that the industry was subject to too much political interference. The owners accepted the return to the Gold Standard as an Act of God which set them a challenge. They were prepared to meet the challenge without help from the Government: a longer working day and a reduction in wages would do the trick very nicely. The miners on the other hand, looked for salvation to an improvement in the industry's efficiency with a subsidy to stave off the necessity for wage reductions. The Government was prepared to buy time by giving a temporary subsidy in order that it might be in an impregnable position during the conflict which was sure to follow; but its longer-term solution was simple: the miner's standard of living had to come down.

2. *Talks with the Owners*

On November 28, 1924, the Mining Association invited the Federation to take part in an investigation of the industry's serious economic position to see what ameliorating steps might be taken. As might be expected however, the two sides discussed the industry's problems on two different levels. For the Federation the talks were, as Mr T. Richards said, a means of putting 'the industry upon a footing that will give them (the miners) better conditions'; whilst for the owners, the important point was to achieve a marked reduction in the cost of production in order to strengthen our competitive position in World markets.

On February 26, 1925, a special Conference of the MFGB was held 'to receive a Report from the Executive Committee upon the Wages Agreement and the present situation in the Mining Industry'. At this Conference most districts put forward their proposals for an amended Wages Agreement. Scotland for example, proposed that the existing Agreement should be abrogated and replaced by a new one designed to restore the miner's real pre-war wage with a reduced working day (six hours as against seven). Most districts thought that some improvement should be demanded. Nottinghamshire however, together with four small districts, had not submitted any proposals. Explaining this, Frank Varley said that whilst Nottingham was not satisfied with the existing agreement, it saw no chance of getting anything better. He went on:

'Now, if I had any criticism to offer on the proposals put before me this morning, it would be this: They are based on a condition of circumstances which actually do not exist. That is all. I want to suggest it is rather an important distinction. Everybody has been addressing themselves to some sort of mythological condition that there is a fund from which we can draw to foot the bill. . . . Nottingham, when the proposal does come, will not be in favour of the termination of the agreement.'

He went on to say that if they did succeed in forcing wages up, then prices would be bound to follow 'because you cannot conceive a condition under which more wages are to be paid from the available wages fund in these districts where the accountants show a loss of $3\frac{1}{2}$d a ton over a long period'. In these circumstances, those owners who were able to make decent profits with present prices would make still higher profits, and Varley was opposed to this. He ended:

'Something has been said in Durham about asking for twelve shillings a day. I wish it could be done, and anything one may say must not be taken to be against the proposals put down in respect of the worth of the miner but from the point of view of being unrealizable. I deplore the holding of a conference like this when the eyes of hundreds of thousands of men are turned towards us, giving the impression that what is put down is likely to materialize when it is not—and in putting down proposals we ought to have some regard . . . (to) what we are likely to attain, having regard to the present position of the industry.'

Mr J. Elks of Kent explained that the reason why his district had sent no resolutions or suggestions to the conference was that they had not so far recovered from the effects of the previous national wage agreement which had led to the unemployment of a majority of his members because the owners were unable to meet the increased wage bill.

In the same debate Frank Hodges, speaking as a delegate of the Midland Federation, appealed for peace in the industry: '. . . a peace which will enable us to get down to the real economic facts and make the necessary readjustments in our industry'.

A. J. Cook, Secretary of the Federation, did not think it was necessary 'at the present time to create an impression of utter hopelessness as must have been the effect of the speeches' of Varley and Hodges. He continued: '. . . We can sum up Mr Varley's speech in these words: "What is the use of anything; why, nothing." As for Mr Hodges' plan for peace, perfect peace, to quote Moleskin Joe, "There's a good time coming by and by, but we may never live to see it".'

At the conclusion of the debate it was agreed that districts should be asked to consider the various proposals further, and that another conference should be called later. At this subsequent conference, held on Wednesday, May 20, 1925, two points of view emerged. The extremists of Lancashire and South Wales wanted the Federation to present a claim for an improved agreement, whilst the Executive favoured a policy of caution as the wording of their motion shows:

'That the report of the investigations into the conditions of the mining industry be accepted and referred to the districts for consideration; also that the Committee continue to make further investigations necessary and report to another conference.'

It was at this conference that A. J. Cook displeased his admirers by his pacific tone which, as we observed in our last chapter, contrasted strangely with his utterances in the country. Replying to his critics, Cook said:

'I may not be as wise as some men, but I am going to be honest in saying—one may change his opinions—regarding tactics. They say a fool never changes, but experience teaches many things, and it has taught me this: if we are going to fight we must have a machine to fight with. I ask you to support the Committee's resolution because I believe the time will come when, despite our having a programme for a living wage, we shall have to meet a national attack upon hours and wages and an attempt to break up the national agreement.'

George Spencer made rather a curious speech at this conference. He asked that the Executive should 'bring another resolution to-morrow morning giving us some definite lead and guidance with regard to this wages question'; and yet, commenting on the Executive Committee's report, he observed that 'we have come up against a set of facts which tell us very plainly that for the moment we can do nothing'. The 1924 Agreement was 'one of the worst ever made so far as some districts were concerned' (and here Spencer was obviously thinking of his own county) but to the poorer counties which had benefited by the provision of a minimum of 33⅓ per cent on standard, it had 'been a godsend'; and for that reason: '. . . the only recommendation that the Executive can bring, and the sooner they bring it the better, is that after considering all the facts, we are thoroughly convinced that this is not the moment to attempt to break up the agreement.'

In the event, the Federation did not need to do any breaking up: the owners did it for them. On June 30th Mr W. A. Lee, secretary of the Mining Association, wrote to A. J. Cook to inform him that the owners had decided to terminate the Agreement of June 18, 1924. Mr Lee expressed the disappointment of the owners: '. . . that the Miners' Federation have not seen their way to agree that the state of the industry calls for an adjustment both of wage-rates and of working conditions.' He went on to say that '. . . the terms which can be offered under a seven-hour day are necessarily much less favourable than those which would be offered under an eight-hour day'; and he made it clear that the Mining Association proposed to hand back to the district colliery owners' associations the negotiating powers they had possessed before the war.

The new terms offered by the owners left the miners gasping. Whereas under the existing agreement wages were a first charge on the industry, the owners not being able to claim a penny until after the minimum wage had been met, under the new proposals the owners would take their proportion of profit whatever misfortune the industry might suffer. The relevant clauses are as follows:

Clause 4: 'In order to determine the percentage payable . . . 87 per cent of the difference between the proceeds in each area and the costs of production other than wages shall be taken; from the amount so determined shall be deducted any special allowances paid under paragraph 5, and the balance so remaining shall be expressed as a percentage of the wages paid at basis rates during the period of ascertainment.'

Clause 5: 'Such provision as may be necessary to meet the case of any low paid day wage man in any district shall be dealt with in that district as a district question.'

It follows from Clauses 4 and 5 that any extra pay given to a very low paid man, over what his base rate would entitle him to, would reduce the fund available for wages generally.

Herbert Smith, at a Special Conference held on Friday, July 3, 1925, called it: 'A very simple form of agreement, an agreement with no bottom in it. . . . What it means is you have 87 per cent after the owners have taken 13 per cent, after costs other than wages have been met to divide between you as you like as to amounts on certain base rates.'

A. J. Cook then explained the effect of the new terms on wages as checked by the auditors. The following reductions in the percentages payable on base rates would apply in August:

In Scotland	A reduction of 47·91 per cent off percentage on base rate.
In Northumberland	A reduction of 47·40 per cent off percentage on base rate.
In Durham	A reduction of 47·66 per cent off percentage on base rate.
In Eastern Area	A reduction of 9·08 per cent off percentage on base rate.
In North Wales	A reduction of 30·14 per cent off percentage on base rate.
In Forest of Dean	A reduction of 33·47 per cent off percentage on base rate.
In Kent	A reduction of 14·18 per cent off percentage on base rate.

Herbert Smith in his Presidential Address warned that:

'They [i.e. the owners] are out to break up any national aspect. I don't want to bring any county in. You see certain proposals made in certain counties. In the event of this policy carrying of getting back to individual districts we should have the use of the words—capacity of the industry to pay. . . . Then we have another subtle thing in the press—some people saying we are not going to have anything to do with a national agreement. What is going to be attempted is this: In certain districts they are going to attempt to say we want no reduction. Well this Federation has got to decide, if it is worth anything, that there has to be no breakaways. There has got to be no individual bargaining, it has got to be done with this Federation or else we split ourselves up into fragments and play into their hands.'

This warning was clearly directed to Nottinghamshire where, as we know, just such an individual settlement was made in 1893. One of the people responsible for that settlement, J. G. Hancock, was still in office and the friendly relations established with the owners a quarter of a century before remained.

However, for the time being there was unity. The conference unanimously rejected the owners' offer on the following grounds:

'1. The removal of the guaranteed minimum wages, which are already below the level of the present cost of living.

'2. The provision of guaranteed profits to the colliery owners, irrespective of the rate of wages.

'3. The immediate great reduction in wages varying from 9·08 to 47·91 per cent on the basic rates.

'4. The continued separation of the mining operations from the profitable undertakings in connection with the coking and by-product departments, etc.'

The Government decided to have yet another Court of Inquiry under the Industrial Courts Act, 1919. This time the Chairman of the Court was the Right Hon H. P. Macmillan, KC, and the other two members were W. Sherwood, Esq. (a Trade Union official) and Sir Josiah Stamp, CBE. The invitation to the MFGB to be represented at the Inquiry was received whilst the Annual Conference was in session at the Grand Hotel, Scarborough. The Executive Committee recommended the conference: 'to inform the Government that it can accept no Court of Inquiry that has for its object the ascertainment of whether mineworkers' wages can be reduced or their hours extended, as these questions were fully discussed at the last Inquiry. The Committee further recommends that we repeat our willingness to meet the coalowners in open conference as soon as they have withdrawn their proposals.'

The view of the Executive was clear: they were prepared to discuss the economic problems of the industry; but only if it were first agreed that there must be no reduction in pay and no increase in hours worked. George Spencer thought that the Executive were making 'an error of judgment'; and he suggested that an attempt should be made to widen the terms of reference of the Inquiry so as to allow a discussion on means of reducing the cost-of-living. He went on to say that, since the miners were asking the General Council of the TUC for support in the dispute, the General Council should have been consulted before a decision on the Government's invitation was reached. He also felt that to ignore the Court of Inquiry would have an unfortunate effect on public opinion; and he ended:

'. . . Suppose notices are not withdrawn and a stoppage takes place, is it suggested there is a possibility of victory within our reach; when we have isolated ourselves and driven away the assistance of the General Council with public opinion up against us? I think therefore this morning, the wisest thing is, before we take this step suggested by the Executive, that more careful consideration should be given to this matter before it is rejected.'

When the Executive Committee's recommendation was put to the vote only two districts voted against—Nottingham and Forest of Dean, the latter because the recommendation was not strong enough.

Undoubtedly the Executive Committee was guilty of an error of judgment on this occasion. In its way, the decision to ignore the Macmillan Inquiry was as bad a mistake as the decision to call out the safety men in 1921. For one thing, the members of the Court were by no means unsympathetic to the miners' case; and for another, the press were able to stigmatize the miners' leaders as unreasonable. Fortunately, Mr Spencer's fear that action was to be taken without consultation with the TUC proved groundless on this occasion.

The TUC General Council had issued a statement on July 10th expressing support for the Miners' Federation and declaring:

'The General Council are confident they will have the backing of the whole organized Trade Union Movement in placing themselves without qualification and unreservedly at the disposal of the Miners' Federation to assist the Federation in any way possible.'

At a special Trades Union Congress held on July 24th to discuss unemployment, the MFGB presented a reasoned statement of the matters in dispute. This was well received. The General Council then proceeded, as Arnot says, to get 'agreement for practical measures of support from the railway and transport unions'.

Meantime, on July 13th, the Court of Inquiry had issued its report which examined the owners' case and found it wanting. Dealing with the proposal that hours should be increased in order to reduce the unit labour cost of output, for example the report says:

'[The argument of the Mining Association involves] a large assumption, namely that such a saving in costs and reduction in price would enable the industry to dispose of its whole increased output, notwithstanding the depressed and disturbed market conditions at home and abroad, and notwithstanding the large stocks of coal known to exist at present in foreign countries.'

The owners were victims of a fallacy: it is true that if one owner could increase the hours of work of his employees without increasing their pay he could reduce his unit cost of production whilst his increased output would have no measurable effect on price. If however, all owners increased the hours of work, then the increased output would be bound to send prices down (assuming an unchanged demand). Further, if we transfer the argument to the international plane, the British mine owners might benefit from increased hours provided that their European competitors agreed to refrain from taking retaliatory action. In practice, this could not be expected to happen. The Mining Association's plan could have no other effect than that of worsening the miners' working conditions without achieving any compensating advantages.

The Court also concluded '. . . that the workers are justified in claiming that any wages agreement which they can be asked to accept should provide for a minimum wage'; and it underlines the fact that: 'The present crisis in the industry, unlike other crises which have arisen in the past, is to a large extent the creation of neither party to the dispute', whilst one member of the Court—Sir Josiah Stamp—went further, as we saw in Section 1 of this Chapter, and allocated a large measure of blame to those responsible for the restoration of the Gold Standard.

3. 'Red' Friday

The lockout notices were due to expire at Midnight on Friday, July 31st. During the preceding week, discussions between the Government and the various parties concerned continued. As late as July 30th, the Prime Minister, Mr Baldwin declared that the Government was 'not prepared to give a subsidy to the industry'. For their part the miners' leaders made it clear that they were not prepared to countenance any reduction in wages or increase in hours. They were strengthened in their attitude by the knowledge that the transport unions had promised to give them full support. The resolutions adopted by the transport unions and endorsed by a special Trades Union Congress begin:

'1. Wagons containing coal must not be attached to any train after midnight on Friday, July 31st, and after this time wagons of coal must not be supplied to any industrial or commercial concerns or put on the tip roads at docks for the coaling of ships.'

They provide for the 'freezing' of all coal movements after Friday midnight in line with the traditional trade union principle of refusing to handle 'black' goods.

The knowledge that on this occasion the Trades Unions meant business produced a change of heart in the Cabinet Council. In the early hours of Friday morning it was announced in a statement issued by the Ministry of Labour, that the Government 'were prepared to render assistance to the industry until the spring'. In the meantime the miners were to continue at work as though the 1924 agreement were still in force; and a full-scale Inquiry into the industry was to take place. Telegrams were sent to all districts announcing:

'Notices suspended. Work as usual—Cook, Secretary.'

The Government had given way, and the militants were inclined to regard this as a victory for drastic action. In practice it was not a victory but a truce; a truce during which the Government intended to prepare for a resumption of hostilities. Mr Hough of Yorkshire, whose views often coincided with those of George Spencer, at a Special Federation Conference held on August 19th thought that:

'. . . it ought to be made clear, that it is no good going about yelling that we have achieved a glorious victory, because we have done no such thing. We have simply retained what we had.

'MR COOK: I want to tell Mr Hough if he does not believe, with the aid of the TUC, that the Miners' Federation have not accomplished a victory he is wrong here—(hear, hear)—I do not think there are many men but what accept the point as I see it.

'MR HOUGH: If you are right, then I am wrong here.

'MR COOK: There is no question about you being wrong. You are never far right in my opinion. I can say without fear or favour that we have achieved a victory, and I am prepared to come to Yorkshire and say the same things there, Mr Hough.'

Mr J. Williams of Forest of Dean, the small Left-wing district which usually contrived to be more Welsh than South Wales, thought that the Executive Committee was wrong in accepting the settlement. He said:

'. . . I think we have had the chance of our lives in this question.

We have had the chance to wrest ourselves free from capitalism, we had more than that, we have had the chance to bring about a real genuine revolution.'

Mr Spencer displayed signs of impatience with this speaker, and then went on to question the platform on Mr Cook's advocacy of revolution about which we wrote above.

The Government as we have said, used the period of truce in late 1925 and early 1926 to prepare for the resumption of hostilities. Early in August 1925, Sir William Joynson Hicks, the Home Secretary, announced at Northampton:

'This thing is not finished. The danger is not over. Sooner or later this question has got to be fought out by the people of this land. Is England to be governed by Parliament and by the Cabinet or by a handful of trade-union leaders?'[4]

There can be no doubt that the vast majority of Trade Union leaders as well as rank and file trade unionists regarded the struggle as an industrial dispute and no more. The miners had been faced with a threat of reduced wages in order to satisfy the ego of those—Treasury pundits and Cabinet jingos—who believed that 'the pound sterling should be made to look the dollar in the face'. Further, the Prime Minister had made it clear that all workers must expect to suffer a reduction in wages 'to put industry on its feet'. For this reason, the miners were able to claim that if the attack on their standards succeeded then all the other classes of workpeople could expect to be attacked in their turn; and for this reason the Trades Union movement rallied to the miners' support. But whilst it is true that for the vast majority of Trade Unionists this was a purely industrial matter, for the small minority this was a 'revolutionary situation' calling for measures designed to sharpen the 'class conflict'. As Mr Francis Williams says, the extreme Left-wing drew the moral that '. . . the power to overthrow the Government by direct action and carry through a social revolution now lay in the hands of the trade unions if only they would have the courage to use it.'

Most of the people who held this view were members of the Communist Party of Great Britain formed in 1920 by the amalgamation of a number of small parties (the principal ones being the British Socialist Party, the Workers' Socialist Federation and the Socialist Labour Party), although there were many syndicalists who remained outside the Communist Party but who, in this period, worked in harmony with the Communists. In 1923 a Miners' Minority Movement had started in South Wales and Scotland, and later in August 1924, this Movement was widened to include workers in all industries.

The President and Secretary of the National Minority Movement—
Tom Mann and Harry Pollitt, respectively—were leaders of the Com-
muninist Party: and the aims of the Movement were stated thus:

'We are not out to disrupt the Unions, or to encourage any new
unions. Our sole object is to unite the workers in the factories by
the formation of factory committees; to work for the formation of
one union for each industry; to strengthen the local Trades
Councils so that they shall be representative of every phase of the
working-class movement, with its roots firmly embedded in the
factories of each locality. We stand for the erection of a real
General Council (of the TUC) that shall have the power to direct,
unite and co-ordinate all the struggles and activities of the trade
unions, and so make it possible to end the present chaos and go
forward in a united attack in order to secure, not only our immediate demands, but to win complete workers' control of
industry.'[5]

The Minority Movement was strongly supported among the more
active members of the NMA, particularly in the Mansfield Area. At
the outset, the chief men in the Movement in Nottinghamshire were
Herbert Booth and G. Williams, a Communist employed at Rufford
Colliery. These two formed a branch after being approached by a
Mr Edgar Davies of Dalby, Yorkshire, and Herbert Booth agreed to
serve on the Executive Committee. However, at the second EC
Meeting, Booth found that the Movement was being run by the
Communist Party and he therefore dropped out. To all intents and
purposes, the Minority Movement was merely an extension of the
Communist Party.

In August 1925, Summit branch tried to get the Association to
affiliate to the Minority Movement but this was rejected. At the same
meeting of Council, C. Bunfield and W. Carter—two extremely
unlikely revolutionaries—were nominated as the Association's
representatives on the local Council of Action set up on the initiative
of the NMA to prepare for the coming industrial struggle.

Mr A. J. Cook continued his week-end revivalist campaign during
which he '. . . increasingly devoted himself' as Francis Williams says,
'to the theme of revolution without being very specific in word or
perhaps in mind, as to what exactly he meant by it'. In relation to the
mining industry Cook said:

'An armistice has been declared, but make no mistake about it,
the issues during the next nine months are far greater than a mere
wage issue. We have got to concentrate our interest on the whole
industry because it is going to be ours.'[6]

And again:

> 'What we did last week (i.e. on "Red Friday") we can and will do again. They have got a commission. Let them sit and see what they can hatch. These enquiries are only held to sidetrack our objective. . . . Before long there will be a political crisis, the greatest in the history of crises, and Labour is going to sweep in in that crisis.'[7]

The Left advocated the perfection of machinery which could be used to run the country during a General Strike, but the General Council of the TUC adopted, as W. H. Crook says, a 'studied attitude of unpreparedness'.[8] Just prior to the 'Red Friday' crisis, a move was afoot to establish a new Industrial Alliance of miners, transport workers, railwaymen and others; but this did not come to fruition largely owing to the opposition of Mr J. H. Thomas of the NUR. Speaking at Chester on August 16, 1925, he said:

> 'What has been accomplished is a truce, but not a permanent peace, brought about by a subsidy. In my considered judgment not only is it wrong but in the end will lead to inevitable disaster in the country.'[9]

Mr Ramsey Macdonald went even further. Speaking at an ILP Summer School at Dunmow he declared that:

> 'The Prime Minister has handed over the appearance, at any rate, of victory to the very forces that sane, well considered, thoroughly well-examined Socialism feels to be probably its greatest enemy.'[10]

Here then we have the picture of a Movement divided; with the extreme Left wing seeking to convert the crisis into a revolution, with the Right wing drawing back in revulsion from the possibility of conflict with the Establishment, and with the bulk of the membership bewildered or apathetic.

On the Government side again there appears to have been a 'war' party determined to bring matters to a head. In December 1925 Mr Churchill made it plain, as Mr Williams points out, that the Government had backed down as it did in July in order to 'postpone the crisis in the hope of averting it or if not of averting it of coping effectually with it when it came'. In order to 'cope with' the crisis the Government divided England and Wales into ten divisions, for each of which a Civil Commissioner was to be appointed with power to take the reins of Government into his hands should an emergency arise. Further, a strike-breaking organization of a somewhat alarming

The Midland Counties
Colliery Owners' Association.

TELEPHONE No 892.
TELEGRAMS- "MINING" DERBY

Regent Buildings, London Road
DERBY.

12th November, 1926.

r Sir,

Wages Agreement for Nottinghamshire.

A Conference has been arranged between representative
tinghamshire Coal Owners and Mr. G.A.Spencer with his
erents representing the men at work at Nottinghamshire
lieries, which will take place on Tuesday, November 16th,
5, at 11.30 a.m., at the Victoria Station Hotel, Nottingham,
g preceded at 10.30 by a meeting of the Owners' representatives.

I am directed to invite the attendance of one
resentative each of the various Colliery Companies having
tinghamshire Pits, and to request that you will intimate the
e of your representative by return of post.

Yours faithfully,

William Saunders.

Secretary.

rs The
igby Colliery Co.Ltd.,
Derby Road,
Nottingham.

3. Letter to coal-owners convening the first wages meeting with Spencer

character, the Organization for the Maintenance of Supplies (OMS), was formed with official blessing.

Mr Williams says that: '. . . what strikes one throughout this period is the total failure of moderate opinion on either side to make any serious attempt to control events. It was as though each were governed by a profound fatalism that enervated both the will to action and the capacity for objective judgment leaving the industrial problems of the time to be carried forward to a climax of violence by forces that because of it assumed the character of historical inevitability.' Had the Miners' Federation had a Robert Smillie for its President or a Thomas Ashton for its Secretary in this period, the course of history might have run more smoothly. In the event, the Miners' Federation leadership had a little too much moral fervour and much too little commonsense. These qualities were to cost the miners, and the nation at large, dear in 1926.

4. *The Samuel Commission*

At the Special Conference of the MFGB held on October 8th, there was a discussion on whether the Federation should recognize the Royal Commission set up by the Government under the chairmanship of Sir Herbert Samuel. Mr Spencer was very insistent that it should. Replying to those who felt that the Commission, like the Macmillan Inquiry, should be ignored, he said that such a decision would have a bad effect upon public opinion and would make the other trades unions much less sympathetic to the miners' cause. However, when the question was first put to the vote, a majority voted against co-operating with the Commission. Despite Mr Spencer, Nottinghamshire joined South Wales, Lancashire, Forest of Dean, Leicestershire, Durham, North Wales and the Cokemen in opposing the Executive Committee's recommendation to appear before the Commission. However, on the following day Conference revoked its decision following an extremely muddled debate.

The members of the 1925 Royal Commission were: the Rt Hon Sir Herbert L. Samuel, Bart (Chairman), Sir William H. Beveridge, KCB, General Sir Herbert A. Lawrence, KCB, and Mr Kenneth Lee, LLD. J. P. Dickie—a biassed observer—says that: 'Except that the miners protested against the absence of any practical miner from the Commission, the appointments met with very little criticism from any quarter'; but it is clear that the composition of the Committee aroused very much deeper feelings than Dickie would admit. The Labour Research Department in its booklet *The Coal Crisis: Facts from the Samuel Commission* (with a foreword by Herbert Smith, President of the MFGB), describes the Commission as '. . . a combination of prosperous individuals who had considerable experience in

K

capitalist government, with prosperous individuals whose financial and commercial interests were a guarantee that they would not betray the capitalist class', and Arnot (who may have had something to do with the writing of *The Coal Crisis*) reinforces this view in his official history of the MFGB.

The Samuel Commission, despite its 'capitalist' composition supported much of the miners' case.[11] They agreed that wages should be settled by National Agreements so as to avoid 'cut throat competition between different districts at the expense of wages'; they agreed that the industry could and should be made more efficient by amalgamations, the closing of inefficient concerns, by 'research into the methods both of mining and using coal', and by the establishment of co-operative selling agencies so as to eliminate wasteful competition. Further, the Commission opposed any lengthening of the working day; supported the proposal that mineral rights should be acquired by the State and agreed that where a mining concern owned coal-using ancillaries, the 'transfer price' of coal used by the ancillaries should be fixed by an impartial authority. This last suggestion was an attempt to meet the argument put up by the Miners' Federation that many concerns were deliberately subsidizing their ancillaries by allowing them to have cheap coal so as to keep miners' wages down. The Commission also agreed that the miners should have some say in the running of the industry through joint committees.

However, upon two major points, the Commission rejected the Federation's submissions. They were opposed to nationalization; and they considered that some immediate reduction in wages was necessary in order to bring the industry's costs into line with its revenue. Further, the Commission opposed any extension of the subsidy: to give a universal subsidy provided profitable concerns with extra profits at the taxpayer's expense; whilst to give a temporary subsidy to marginal producers 'would constitute in many cases a dole to the inefficient to the disadvantage of the efficient'.

Arnot believes that 'The parties to the Report wished to divide the Labour Movement, to convince those who were ready to be convinced as well as those who were less ready, that the miners should suffer a reduction in wages and that the other trade unions would not be justified in repeating their action of the previous July'. Another Communist writer, Mr Allen Hutt, expresses a similar view. The Samuel Commission produced says Hutt, 'a Report which was vague in its references to State intervention for the reorganizing of coal capitalism, but precise in its assertions that the miners should accept longer hours or lower wages'. (In fact, the Commission made it clear that wage reductions should be conditional on the introduction of measures of reorganization. The passage reads: 'Before any sacrifices are asked from those engaged in the industry, it shall be definitely agreed between them that all practicable means for improving its organization and increasing its efficiency should be adopted as speedily as the circumstances in each case allow.')

It would have been perfectly natural for partisans like Mr Arnot and Mr Hutt to express the opinions quoted above in the heat of the moment, but one finds it difficult to understand how anyone who has had a quarter of a century or more to think things over could write off the Samuel Commission so cheaply. The Commission's Report, which was subsequently endorsed by the Government, contained much that was useful; but in the circumstances of 1926 it fell on deaf ears. The MFGB insisted that wages should not be reduced; the owners insisted that reductions were necessary. Further, whilst the Federation refused to countenance district negotiations, the owners refused to negotiate nationally; although in view of the Government's attitude, this particular stumbling block could have been got over had there been any possibility of reaching a settlement on the question of wages.

During this period of uneasy truce, the Left-wing branches of the NMA kept up their agitation for one-hundred-per-cent membership; but with scant success. Despite the mammoth rallies addressed by A. J. Cook and others, membership showed no signs of increasing to any extent. Because of this, a number of lodges: Bulwell, Huthwaite, Welbeck, Bentinck and Annesley, sought permission to strike in order to force the non-members into the Association but this, again, was without result.

The Summit Lodge went so far as to advocate a General Strike in January 1926 to secure the release of the Communists who had been imprisoned as Hutt says, 'after the biggest state trial since Chartist days', and this motion was endorsed by Council.

Other motions submitted by various Lodges wear an air of unreality when one considers the industry's position at that time. Linby appealed for the enforcement of the all-throw-in system; Cinder Hill asked that the Joint Board (with the Derbyshire Miners' Association and the Enginemen and Firemen's Union) should take steps to enforce the demand for the restoration of the $7\frac{1}{2}$ per cent cut in non-Top Hard wage rates; whilst on March 27, 1926, Rufford asked that a Resolution should be placed on the agenda of the Annual Conference of the MFGB asking that wages should be based on the cost of living (presumably this means that the 1914 wage rates should be scaled up in accordance with the official cost of living index); that the working day should be reduced to six hours, that there should be an annual week's holiday with pay, and that the mines should be nationalized with workers' control. Like Nelson, these delegates viewed the World around them through a blinded eye. The once-strong Nottinghamshire Miners' Association might still have the appearance of solidarity, but it was now little more than a paper tiger. Of its thirty-three thousand members, no fewer than six thousand were in arrears with their subscriptions and it was not until March 1926 that the huge debt accumulated by the Association in the 1921 Lockout was liquidated. Some ten thousand men stood outside the union altogether, whilst those who were in had the unedifying spectacle of a moderate district leadership being prodded into activity by a Left-wing 'ginger group' ably abetted by the secretary of the national Federation.

APPENDIX

RATES OF WAGES AND HOURS OF LABOUR OF CERTAIN CLASSES OF WORKPEOPLE IN 1925*

	Pay (per wk.)	Weekly Hours
i. LOCAL AUTHORITY EMPLOYEES (NOTTINGHAM CORPORATION)†	s. d.	
General Yard Labourers	54 10	47
Road Labourers	54 10	47
Scavengers	54 10	47
Refuse Collectors: Horse Drivers	56 0	48
Motor Drivers	63 0	48
Loaders	57 0	48
ii. COAL TRIMMERS		
Employed on Tyne Docks‡	117 0	
Employed at Cardiff, Penarth & Barry §	84 10	
iii. RAILWAY SERVICE†		
Goods Porters (Industrial Areas)	50 0	48
Porters, Grade II (Industrial Areas)	46 0	48
Engine Cleaners	46 4	48
iv. BUILDING—LABOURERS†	55 7	44¼
v. ELECTRICITY SUPPLY—LABOURERS†	54 10	47
vi. GAS WORKS—LABOURERS†	52 11	47
vii. MINING		
Average Weekly Earnings of Miners in Eastern Division:		
A—First Quarter 1925	56 9	42
B—Second Quarter 1925	49 2	42

* *Sources:* Report of Samuel Commission 1925, Vol. 1 (Cmd. 2600) pp. 156–161 and *The Coal Crisis: Facts from the Samuel Commission 1925–26* (L.R.D.), p. 60.

† Standard Rates.

‡ Average for first nine months of 1925.

§ Average for first eleven months of 1925.

CHAPTER 8

THE GENERAL STRIKE

1. *Introduction*

The story of the General Strike has been told so often that there is no need for us to go into great detail: a brief account will suffice.[1]

The Industrial Committee of the TUC and the negotiating committee of the MFGB met on February 19, 1926 and subsequently issued a statement which said:

> 'The attitude of the Trade Union Movement was made perfectly clear last July, namely, that it would stand firmly and unitedly against any attempt further to degrade the standard of life in the coalfields. There was to be no reduction in the wages, no increase in working hours, and no interference with the principle of National Agreements. This is the position of the Trade Union Movement today.'

On March 12th, a special conference of the MFGB was held to consider the Samuel Report which had been published two days earlier. Conference decided to refer the matter back to districts in order that the membership could have a chance to study the Report before expressing their opinions. George Spencer was worried about the effect which week-end speeches might have upon the miners. He asked: 'Are some men going to assume the right to go on the platform this week-end and not only explain the [Samuel] recommendations, but interpret and give their personal views. If the South Wales leaders, or any other leaders are going to express their own personal point of view, I say that is bound to have a prejudicial effect on this question.'

On March 24th, the Prime Minister announced that '. . . if the other parties accept the recommendations of that Report [i.e. the Samuel Report] I do'. However, the discussions which took place between the 'other parties'—the Mining Association and the Miners' Federation—proved fruitless; so much so that it was obvious that, left to themselves, the two would never agree. The Mining Association still insisted not merely on reductions in wages, but also on increased hours. The Federation on the other hand, refused to consider any possibility of longer hours or less pay. On the question of national as against district negotiations, the owners reluctantly accepted the

150

Samuel Commission's recommendation, and on April 1st they proposed:

> '. . . that the two sides should meet at once to formulate the terms of the proposed National Wages Agreement, and that meetings should be held forthwith in the various districts to deal with the matters that call for the attention of the district associations, the most urgent of which is the settlement of the amount of the minimum percentage on basis rates and the amount of the subsistence wage in each district.'

The owners agreed 'in deference to the Commission's view' that the general principles of a new agreement should be settled nationally. These general principles should include: 'the ratio between profits and wages; the details of the method of ascertainment and the definition of proceeds and of the items of costs of production.' The owners further agreed that the 'amounts of the minimum percentages settled in the various districts should be submitted to the national conference for approval'. This did not quite meet the Federation's view on this issue (since the minimum percentages were to be fixed in the districts and ratified at national level instead of there being one minimum percentage applicable to the whole country and negotiated nationally); but it went a long way towards it. However, the MFGB Conference held on April 9th recommended districts to insist upon 'a national wage agreement with a national minimum percentage'; no wage reductions and no increase in hours. The Nottinghamshire delegation at this Conference was however, uneasy. One delegate asked: 'When I get back to Nottinghamshire, in all probability the branch I represent when I state the recommendation of the Executive Committee, they will want to know as to what are the chances, what are the odds of obtaining them. That is the great question for Nottinghamshire and I want to know what are the betting chances?' This question was not properly answered, Herbert Smith refusing to act as tipster. Later in the debate, George Spencer came out strongly against the Executive's recommendation. He thought the Samuel Report contained 'elements of usefulness if only they could be applied'. Further, the Government had promised to continue the subsidy for a temporary period, and Spencer thought that the Federation should explore this a little more fully to see whether the additional subsidy could help the industry over its reorganization period. He was sure that if the Federation were simply to refuse to negotiate further on this issue then public opinion would turn against them. Spencer's view on this question did not attract any support however, and Conference agreed to endorse the Executive Committee's recommendation to districts, which reads:

'That this Conference having considered the Report of the Royal Commission and the proposals of the colliery owners thereon, recommends the districts as follows:

'(*a*) That no assent be given for any proposal for increasing the length of the working day.

'(*b*) That the principles of a National Wage Agreement with a national minimum percentage be firmly adhered to.

'(*c*) That inasmuch as wages are already too low we cannot assent to any proposals for reducing wages. These recommendations to be remitted to the districts for their immediate consideration and decision, after which a further conference be called as speedily as possible for the purpose of arriving at a final decision.'

In the middle of April the owners in most districts gave notice to terminate contracts at the pitheads. In Nottinghamshire however, lockout notices were not posted and Council therefore decided to hand in strike notices. The wording of the resolution on this subject, adopted on April 20, 1926, is suggestive:

'Resolved that after full consideration of *several suggested alternatives*, and with a view to keeping faith with the Federation, we do instruct all Secretaries to give in a collective notice to cease work on April 30th at every colliery in the Nottinghamshire Miners' Association, same to be signed by the Branch Secretary.'

One wonders what the 'several suggested alternatives' were. Presumably George Spencer carried the arguments he had used at national conferences into the council chamber. However that may be, Council also resolved: 'that we do not see our way clear to accept Employers' invitation to a joint conference to discuss the question of the division of the Eastern Area seeing that Federation Executive has decided that districts shall not approach Employers individually on the question.' At the next meeting of Council, held on April 24th, it was resolved: '. . . that this Council instructs the whole of its branches to make the necessary preparations to cease work at the end of the afternoon shift on Friday, April 30th next, in the event of settlement or suspension other instructions to be issued.' The Nottinghamshire miners were once more to 'come out for nothing' to keep faith with the national Federation. In 1893, when the dispute was with the coalowners only, the Association was able to negotiate a return to work with the Leen Valley owners and the men who went in to work helped to support those who were still out. Now however, as in 1921, the dispute was with not merely the owners but also the Government, and local settlements were not therefore to be allowed.

In 1893 the Federation won by denying the owners their profits: those owners who cared to pay the pre-stoppage rates of wages were allowed to re-engage their men, and they could then make hay at the expense of the other owners who insisted that a reduction was necessary. In 1926 however, the Federation hoped to win by denying the nation its coal: for this reason those who produced or carried coal were regarded as 'blacklegs' no matter how high may have been their remuneration. In the event, the Federation's stranglehold was broken; and the men were to be driven back under conditions tantamount to unconditional surrender.

By the middle of April the owners had gone back to their insistence on purely District settlements: they insisted that the Mining Association had no power to negotiate nationally. The various district owners' associations had published the terms on which the pits could continue at work after May 1st: providing in many cases for reductions so large as to make their rejection inevitable. The average reductions were estimated to be: 2s 1d a day in Scotland; 2s 4d in Northumberland; 2s 9d in Durham; 2s 10d in South Wales; 1s 0d in Yorkshire; 1s 7d in Lancashire; 1s 8d in North Wales; 1s 7d in South Staffordshire and Shropshire; 2s 7d in Cumberland and 1s 3d in the Forest of Dean.[2] The Government undertook to try to persuade the owners to continue national negotiations resulting in a national minimum wage and eventually, on April 30th, they succeeded. However, the prior conditions on which the owners insisted—the acceptance of an eight-hour day and a cut in wages—rendered this success illusory.

By this time the General Council of the TUC had taken a hand in the discussions. The Trade Union Movement as a whole saw the attack on the living standards of the miners as the prelude to an orgy of wage-cutting; and for this reason, the various Unions had empowered the General Council to render the miners every assistance in their struggle. On April 29th, the General Council summoned the Executive Committees of the affiliated organisations to a special conference in London. Next day the following letter was despatched to the Prime Minister:

'To the Rt Hon Stanley Baldwin, MP.

'Dear Sir,

'I have to advise you that the Executive Committees of the Trades Union Congress, including the Miners' Federation of Great Britain, have decided to hand over to the General Council of the Trades Union Congress the conduct of the dispute, and all the negotiations therewith will be undertaken by the General Council. I am directed to say that the General Council will hold themselves available at any moment should the Government desire to discuss the matter further.

'Yours faithfully,

(signed) WALTER M. CITRINE,
"Acting Secretary".'[3]

On the same day the Government declared a state of emergency. The names of the Civil Commissioners and their staffs were published; placards appealing for volunteers for the Organization for the Maintenance of Supplies were posted all over the country and ostentatious troop movements took place.

On the evening of the following day, May 1st, a sub-committee of the General Council met the Prime Minister in an endeavour to effect a compromise; and the negotiations continued on May 2nd. Eventually, the following formula emerged:

'The Prime Minister has satisfied himself as a result of the conversations he has had with the representatives of the TUC that, if negotiations are continued (it being understood that the notices cease to be operative), the representatives of the TUC are confident that a settlement can be reached on the lines of the [Samuel] report within a fortnight.'[4]

The Executive Committee of the MFGB were annoyed at the action of the General Council in entering into further negotiations. This emerges from the Report prepared by A. J. Cook, the relevant paragraph of which reads:

'The action of the TUC representatives in communicating with the Government and agreeing with them upon the formula was taken on their own responsibility and without prior consultation with the representatives of the miners. In fact it was only after these events had taken place that the Secretary of the Federation learned of them unofficially, and telephoned to the Secretary of the Trades Union Congress for confirmation. The Secretary was then called before the full General Council and informed by it that it had met the Government and accepted the formula. The Secretary in the name of the Miners' Federation, protested at the irregularity of the proceedings.'

Despite the protest of the Miners' Federation, the General Council was determined to avoid a conflict. Unfortunately their discussions with the Prime Minister were broken off by the latter owing to the action taken by printers employed by the *Daily Mail*, who had refused to print the paper unless a leading article bearing the headline, 'For King and Country' were withdrawn. The Prime Minister refused to continue the discussions unless the General Strike notices, which had been sent out on May 1st, were unconditionally retracted. This demand precipitated the very conflict which it ostensibly sought to avoid; and it is difficult to resist the feeling that the militaristic members of the Cabinet were determined to crush the serpent of militant Trade Unionism once for all. The extremists of both sides were to be indulged with the open class warfare for which they had waited since the ending of the Great War.

2. *The Nine Days*

The Conference of Trade Union Executives which met on April 29th and the two succeeding days, took the decision to organize a General strike in defence of the miners' standard of living, should it prove impossible to effect a satisfactory settlement. The General Council's principal spokesmen, J. H. Thomas and Ernest Bevin belonged to the Movement's right wing and they made it clear in their speeches that they were being driven to do what they did. J. H. Thomas said:

'My friends, when the verbatim reports are written, I suppose my usual critics will say that Thomas was almost grovelling, and it is true. In all my long experience—and I have conducted many negotiations— I say to you—and my colleagues will bear testimony to it—I never begged and pleaded like I begged and pleaded all day today, and I pleaded not alone because I believed in the case of the miners, but because I believed in my bones that my duty to the country involved it. . . . But we failed. The Cabinet was summoned—such additional members as had not previously been in the negotiations, a number of whom had heard nothing of what had taken place, were called in to give their final decision, and their decision was a refusal to accede to your request. Please observe, not effect a settlement, but a refusal to accede to your request for a suspension of the notices so that negotiations could continue.'

Mr Bevin, in introducing to the conference the General Council's plans for organizing the General Strike, apologized for its being done in a hurry, since the General Council had believed until the last moment that 'peace would accrue'. He compared the Government's action in 'mobilizing the forces of war' with that of the Government of Lord North and George III whose belligerent attitude precipitated the American War of Independence and he warned that 'the result may be fraught with as serious consequences as the action of George III in the history of this country'.

The General Council's plans for co-ordinated strike action specified those trades which were to come out: transport workers; printers; iron and steel trades, metal and heavy chemicals workers; building workers (except those engaged exclusively on building houses); and electricity and gas operatives. On the other hand, the constitutent unions were charged with responsibility for ensuring the distribution of food-stuffs; and were to see that the needs of hospitals, convalescent homes, sanatoria and schools were met.

The trades councils, in co-operation with the local officers of the unions actually involved, were instructed to assist in carrying out the TUC's policy on the maintenance of essential services; and were to organize 'the Trade Unionists in dispute in the most effective manner for the preservation of peace and order'. The General Council also warned trade unionists to beware of agents-provocateurs, who were to be 'dealt with immediately'.

On Tuesday, May 4th, the transport workers, the printers and others, stayed at home. The General Council announced that the difficulty was not to persuade men to come out, but rather to persuade those who were in the 'second line of defence' to keep going to work. In the gas and electricity industries however, the supervisory staffs were able to maintain a service despite the absence of the operatives.

In Nottingham, the strike arrangements were the responsibility of a Joint Advisory Dispute Committee which comprised the Executive Committee of the Nottingham Trades Council, two representatives from each union affected, or likely to be affected by the strike, and representatives from the Long Eaton and Netherfield Dispute Committees.

The main committee, and its sub-committees (Permits; Publicity; Meetings; and Outside Pickets) met daily. The Permits Sub-Committee 'considered and endorsed permits for the removal of foodstuffs issued by the Joint Transport Strike Committee'.[5] The Publicity Sub-Committee issued a daily bulletin and they were helped in this work by 'comrades of the ILP'. The Meetings Sub-Committee arranged meetings 'every day in all parts of the city and surrounding districts, served by a rota of speakers'. These meetings were 'of valuable assistance in maintaining the morale of the men'. Contact was maintained with London by the use of dispatch riders and deputations were exchanged with neighbouring towns.

On May 4th the Nottingham Corporation 'bus and tram services were idle. The few private 'buses which ventured into the town were attacked and overturned by strikers. During a demonstration in Old Market Square, the private car belonging to a local 'bus proprietor, Mr A. Barton, whose employees were non-unionists, was also overturned. On the same day some 2,000 demonstrators toured the Queens Road area and demanded that work should stop at a number of local factories. This was quite clearly contrary to the instructions of the Strike Committee but a number of works, including an engineering factory, a lace factory and a timber yard, were brought to a standstill. At nine o'clock that evening some 400 demonstrators, supposedly led by 'Seaman' Carrington, a local boxer, besieged the Palais-de-dance, but a posse of policemen with drawn truncheons forced them to disperse.[6]

After the first day however, the Strike Committee managed to assert its authority and acts of violence were few.

There was a similar committee at Mansfield, but this does not appear to have functioned at all well. The Trades Council reported later that 'The occasion was too big for the local machinery to deal with'; and although a daily strike bulletin was prepared, only two dozen copies were issued. The Mansfield District Committee whose secretary at this time was Jack Simms, a Derbyshire Deputy, took part in the work of organizing the workers, but its activities were far more effective during the prolonged mining dispute which followed the Strike.

In some parts of the country however, the Disputes Committees were remarkably effective and exercised considerable power over, for example, the distribution of foodstuffs. At the same time the Government's emergency arrangements worked well, as Francis Williams acknowledges.

The TUC General Council was desperately anxious to avoid disorder; and in Lincoln the Trades Council supplied the authorities with the whole of the Special Constabulory force. However, on the Government side, the desire to 'teach the trade unions a lesson' stood in the way of an early settlement. Indeed, the Archbishop of Canterbury's proposals for a compromise solution of the dispute were deliberately ignored by the Government paper, *The British Gazette* (which was supposedly edited by Mr—now Sir—Winston Churchill).[7] The Government also brought pressure on the BBC to suppress the Archbishop's appeal; and it was left to the TUCs strike broadsheet: *The British Worker* to give publicity to it.

However, if the Government was, as Mr Francis Williams asserts on the authority of Earl Winterton's memoirs, 'under heavy pressure from many sides to do nothing which might be regarded as in any way conciliatory but instead to force "a showdown",' the leaders of the TUC on the other hand, were searching desperately for a straw to clutch. The straw was provided by Sir Herbert Samuel who, after consultation with the parties concerned (Government, colliery owners, Trades Union Congress General Council and the MFGB Executive) presented a draft memorandum embodying proposed terms of settlement. These terms followed closely the recommendations of the Royal Commission, of which Sir Herbert had been Chairman, the main provisions being: a renewal of the subsidy whilst negotiations on miners' wages continued; the establishment of a National Wages Board with miners' and owners' representatives, plus a neutral element and an independent chairman; 'no revision of the previous wage rates unless there are sufficient assurances that the measures of reorganization proposed by the Commission will be effectively adopted';[8] any wage agreement drawn up 'should not adversely affect in any way the wages of the lowest paid men'[9] and should fix for every class of labour a firm minimum wage; a ban on the recruitment of inexperienced adult labour whilst experienced miners remained unemployed; and provisions for assisting men rendered redundant and those required to seek jobs in other areas owing to the proposed reorganization of the industry.

This memorandum was not an official document as Sir Herbert himself made clear. For all that, the TUC General Council were of the opinion that it offered a reasonable basis on which a settlement could be arranged and because of this they decided to call off the strike.

Accordingly, on May 12th, the General Council met the Prime Minister to acquaint him of their decision. The Prime Minister welcomed their decision and promised to 'use every endeavour . . . to ensure a just and lasting settlement'.

The next day *The British Gazette*, as Mr Williams says, 'exultantly proclaimed the complete capitulation of the trade unions and announced that the Government's victory had been secured without giving any guarantees regarding the re-instatement of strikers, the renewal of trade union agreements, or the withdrawal of mining Lockout notices'. This attitude communicated itself to employers, many of whom sought to victimize their employees. The railway companies for example, announced that 'all strikers were considered as dismissed and would only be re-employed on individual contracts'. For two days 'almost as many men remained out as at the height of the stoppage' because of the harsh conditions which employers sought to impose, and a very real danger of a resumption of hostilities—this time with the left-wing militants in charge—existed.

In this situation Mr Baldwin who, for all his faults had a great deal more common sense than some of his more active colleagues, 'told the House of Commons in the strongest terms that the Government could not countenance any attempt by employers to use the end of the strike to enforce wage-cuts or alter the terms of existing agreements'. Following on this, the employers' and workers' organizations arranged terms for an organized resumption of work. However, in the mining industry the men remained out. The MFGB refused for the time being to accept any solution which provided for reduced pay or increased hours. The leaders of the Federation also felt that they had been betrayed by their colleagues in other trades, and this view was accepted widely by the men. This view was supported too, by the Amalgamated Society of Woodworkers, whose Executive Committee characterized the action of the TUC General Council in calling off the strike as an 'inglorious and humiliating capitulation . . . one of the most deplorable and discreditable episodes in the history of trade unionism'.[10] The ASW Executive sent a protest to the TUC, with copies to all other affiliated unions. This reads:

'May 13th, 1926.

'Mr W. Citrine, Acting Secretary,
  'Trades Union Congress General Council.

'Dear Sir,

  'I give below resolution passed by my Executive Council at its meeting held this day, and I have to request that it has the immediate consideration of the Trades Union Congress General Council.

'Yours faithfully,

(signed) F. WOLSTENCROFT,

'General Secretary.

  'That this Executive Council now in session views with grave concern the instruction received from the Trades Union Congress that the General Strike initiated by that body with the full approval of the affiliated unions has been called off without the consideration and consent of the unions concerned. It therefore demands, having regard to the future power and well-being of the trade union movement in this country, that a special conference of all executive councils be immediately convened to enable the General Council to report on the whole proceedings and to devise plans for the future.'

For the General Council of the TUC however, there was no future to consider in relation to general strike action. Mr C. T. Cramp of the National Union of Railwaymen was doubtless expressing the common view when he declared 'Never Again!' The General Strike had demonstrated that militancy, taken to its logical conclusion, entailed a willingness to take over the reins of authority from Government and Parliament. To the moderate trade union and labour leaders this was as Mr Arthur Henderson said, a 'terrible prospect'. The leaders of the TUC and Labour Party had refused to prepare for the Strike: as Mr Kingsley Martin, writing in 1927, pointed out: 'No workers' defence corps was formed, no "factory committees" were appointed, no organization for a rival service to that of the Government was countenanced, although all these measure were loudly demanded by Communists. On the contrary, the TUC offered to aid in the maintenance of food supplies and gave incessant injunctions to the strikers to act in an orderly, peaceful and legal manner. The TUC never attempted nor intended to attempt to act as a "rival Government".' For all that: '. . . the sympathetic strike offers an opportunity to those who advocate a violent settlement of the class-war on both sides. As Mr Clynes has said, you cannot call a general strike without the knowledge that it may lead in time to a civil war.'[11]

But if the moderates regarded the General Strike as an experiment too dangerous to bear repetition, there were those on labour's left who felt that the experiences of 1926 should be used to ensure a more successful effort on any future occasion. This view is well expressed by Mr Emile Burns who says of his book *The General Strike, May, 1926: Trades Councils in Action*:

'But it is not merely as a historical record that the book has been prepared. The General Strike of May 1926 cannot be regarded as an isolated event; "Never Again" is the despairing prayer of individuals who do not like the course of events, rather than a serious judgment of what the course of events is likely to be. No one in the working-class movement can fail to realize that the conditions which led up to the General Strike are still in existence, and that they will continue so long as the capitalist organization of society exists. The strike even of a single union on a national scale is a comparatively modern feature; the widening of the conflict in the development of the class struggle is a simple historical fact.

'Therefore the study of the local organization and activities during the General Strike of May 1926, is important in order that the working-class movement as a whole, and in particular the local organizations, may be able to learn the lessons of experience, and make the necessary preparations to deal effectively with any similar emergencies which may arise in the future.'

No doubt this left-wing point of view found a large measure of support among the rank-and-file; but the leadership of the movement turned their back on 'direct action'. They recognized that the immense damage caused to the economy by a General Strike should be avoided at all costs. On the other side, many employers had also learnt a lesson. They recognized that the Trade Unions possessed real power which could not easily be broken. They recognized further that standing behind the moderate leaders of the TUC were other men, less moderate, who were willing to carry 'direct action' to its logical conclusion given the right conditions. For such employers there were two possible lines of action: they could either foster 'client' trade union leaders in the hope that such people would be able to wean

Agreement

MADE BETWEEN THE

NOTTINGHAMSHIRE & DISTRICT COLLIERY OWNERS

AND

REPRESENTATIVES OF THE WORKMEN

working at Nottinghamshire and District Collieries.

November 20th, 1926.

Eustace Mitton

4. G. A. Spencer's copy of the 'Spencer' agreement autographed by H. Eustace Mitton, Chairman of the Midland Counties Colliery Owners Association

the workers away from the old unions with their socialist affiliations, or they could achieve a *modus vivendi* with the official trade union leadership, assuming that the latter would be able and willing to keep their supporters in a peaceable frame of mind. In a later chapter we shall see that whilst the first alternative (which went by the name of 'non-political trade unionism') was tried, it achieved little success outside the Nottinghamshire coalfield; and we shall argue that this was due to the alacrity with which the TUC took up the idea of co-operation with the employers in the rationalization of industry—a policy which has come to be known as Mondism. Whilst the official trade union movement maintained its political affiliation, it was to insist increasingly on the rigid division between industrial and political matters, the one to be settled between employers and employed, and the other to be dealt with in Parliament. Indeed, it is hardly going too far to suggest that the TUC, whilst condemning Spencer as a schismatic, accepted the fundamentals of his philosophy. Mondism was Spencerism in respectable attire.

CHAPTER 9

THE 1926 LOCKOUT

1. *The Opening Stages*

The General Strike was over, but the miners' lockout was only just beginning. The MFGB Conference on May 15th received an outline of proposals for the settlement of the dispute from the Prime Minister. These proposals followed closely the recommendations of the Samuel Report. They provided for the introduction of bills to secure amalgamations of mining companies, the creation of a welfare levy on royalty owners, restriction of recruitment and a National Wages Board on the lines of the Railway Wages Board. The Government also undertook to set up various committees (e.g. one to 'investigate the question of selling syndicates', and another to 'examine the profit-sharing proposals of the Commission and family allowances') and to prepare a scheme for the establishment of Pit Committees with workers' representation. The Government also agreed to provide a further subvention in aid of wages of approximately £3 millions, but it expected the miners to accept a reduction in minimum wage-rates (excluding subsistence rates) in all districts. A national board with three members from each side of the industry and an independent chairman would be set up to 'frame a National Wages and Hours Agreement governing the principles on which the general wage-rates should be ascertained in each district', and to decide what the districts should be. Where the two sides were unable to agree on any point, the independent chairman would have power to settle. Item 7 of the proposals reads:

'If the parties agree that it is advisable that some temporary modification should be made in the statutory hours of work, the Government will propose the necessary legislation forthwith and find facilities for its immediate passage.'

Conference adjourned until May 20th in order that districts might mandate their delegates. When the delegates reassembled they were unanimous in rejecting the Government's proposals although Mr F. Swift of Somerset warned that the miners were likely to be in a weaker bargaining position the longer they held out. He moved an amendment which was later withdrawn, which sought to authorize the Executive to continue negotiations with the Government in order to reach a settlement. The attitude of Nottinghamshire was much the same. At the Council meeting of May 17th it was resolved that the Executive of the Federation should be empowered to negotiate terms for a settlement and that: '. . . the said terms shall not necessarily follow either the Government or the Samuel proposals but shall be submitted to a Conference and to the members for their acceptance or otherwise.'

Nottinghamshire's anxiety to bring the stoppage to an early conclusion was partly due, no doubt, to the serious financial difficulty in which it found itself. The Association entered the dispute practically penniless, the debts incurred during the 1921 lockout having been liquidated only a short time before. At the Federation Conference held on May 20th, George Spencer suggested that the poorer districts like Nottinghamshire should receive a larger share of the funds received by the MFGB from the public than the better-off districts. Mr Spencer's appeal was rejected by Conference. However, the Derbyshire Miners' Association, despite their own difficulties, made a gift of £10,000 to the NMA towards the end of May 1926, and this together with the £7,500 received from the central relief fund enabled the union to pay its members a week's strike pay. Thereafter, the Association relied for its income almost entirely upon the periodical payments from the central relief fund to which the Russian trade unions contributed so very largely. Even so, members received no more than a few shillings a month. For those members who were unable or unwilling to share in the profitable business of outcropping, suffering was intense.

Unfortunately, the Association's credit now stood at a very low ebb. Those traders who had accepted coupons for goods in 1921 and had been kept waiting so long for their money were unwilling to repeat the experiment; and whilst the union appealed to the general public for funds the result was not nearly so gratifying as in 1893, partly on account of the general suffering and partly on account of the much smaller measure of sympathy which the middle classes had for the miners in the quasi-revolutionary situation of the 1920s.

Right from the start Nottingham appeared as the weak link in the MFGB chain. As early as Saturday, May 1st at a Federation conference, George Spencer had expressed his opposition to a stoppage: there were he said, people in the House of Commons who were trying to prove that the Federation's policy was being determined by a small pressure Group and not by the rank-and-file. He continued: 'I remember very distinctly and everybody else does, that some of the men in 1921 at the commencement of the stoppage, were loudest in their approval of that stoppage, were the very men who were loudest in their condemnation of the Executive and everybody else when it was a failure, and said "We ought not to have decided, why not a ballot and why did you not do this and that?" You have had warning. There is the same measure of enthusiasm amongst a comparative number of representatives here and outside, and if it is a failure they will say they did not function. . . . We are going to be told again that we have taken this step without consulting the rank-and-file.'

Spencer claimed to be speaking on behalf of his county, but William Carter did not accept Spencer's interpretation of Nottinghamshire's attitude. 'I say as the General Secretary of the Nottinghamshire Miners' Association,' he said, 'it would be a crime and a calamity if any man went out of this conference with the idea that Nottingham-shire were not true; we have always been true, and will be true and will not depart from the agreed policy of the Miners' Federation of Great Britain.'

Again in April 1926 Frank Hodges, A. J. Cook's predecessor who was now secretary of the Miners' International, speaking at Notting-ham had attacked the policy of the Federation and said 'that the miners would be prepared to accept longer hours in preference to lower wages'.[1] Hodges was originally regarded as belonging to the left-wing: he was indeed one of the students at Ruskin College in 1908 who revolted against the teaching of 'capitalist' economics and who formed a rival institution—the Central Labour College—to teach Marxian economics. Later however, he moved further and further to the right finishing up as a wealthy company director. By 1926 he had already become a rallying point for those who opposed the policies of the predominantly left-wing leadership of the Federa-tion, and his appearance in Nottingham on the eve of the lockout was not without significance.

A month later, on May 28th, Frank Varley who had presented the MFGB's evidence on wages to the Samuel Commission, proposed a compromise solution providing for a 'reduced national minimum percentage till the end of the year, and a continuance of the subsidy to cover losses only'.[2] The great drawback to the provision of a subsidy in the ordinary way is that the prosperous producers receive an unearned and unnecessary bonus at the expense of the country at large. Varley's proposal would have reduced the burden placed on the national economy by a continuation of the subsidy by restricting payments to the sub-marginal producers. The prime objection to this idea however, is that it destroys—or at least reduces—the incentive to efficiency. Such considerations are purely academic since the MFGB rejected the proposals and condemned Varley for making them. A number of branches of the NMA sent in strongly-worded motions to Council: Wollaton asked that Varley and Spencer should take no further part in the negotiations; Welbeck and Sherwood wanted the Association to censure Varley whilst Newcastle and Bentinck felt that he should be given three months' notice to terminate his contract. However, none of these motions received Council's endorsement.

Not only was there weakness at the top, but there was also weakness among the men. In Nottinghamshire an unusually large number of men worked outcrop coal right from the beginning of the dispute. The NMA Council on July 5th appealed to the Mines Inspectorate, the police and the local councils to stop outcropping, since its own efforts to discipline its members had had no effect. Moreover, at the new Blidworth and Clipstone Collieries, where coal had just been reached, work continued throughout the lockout. Meetings and demonstrations were held at Blidworth—one as early as May 30th— to try to induce the men to cease work, but to no avail. The curious thing about this is that some of the permanent officials of the NMA had agreed to permit these two pits to work since they were developing pits. There was then, weakness at the top and at the bottom; and this weakness was probed both by the owners and by the authorities. The owners tried almost continuously to induce a breakaway from the national movement, and in this they were helped by the fact that the employment contracts in Nottinghamshire were terminated not by them but by the Union: the men could, had they wanted, have remained at work on the old rates of wages. So far as the authorities were concerned, many people who played an active part in the dispute insist that in Nottinghamshire and Derbyshire the police went out of their way to make life uncomfortable to the men. Pickets were intimidated and assaulted, and in odd cases, arrested. The present writer's view is that in practice the right to picket peacefully is subject to severe limitations: the well-known impartiality of the British constabulary does not necessarily extend to trade disputes. It appears however, that in this particular struggle partiality was carried to a fine art. Far from keeping the peace, the police only too often provoked breaches of it.

2. The 'Bishop's Proposals'

At the meeting of the MFGB Executive Committee held on June 29th, when the dispute was two months old, a letter was read from the Industrial Christian Fellowship, an inter-denominational organization. The Fellowship took the view that the solution of the mining dispute was to be sought in the adoption of the Samuel Commission's findings and they asked that the Federation should agree to appoint a small number of representatives to meet representatives of the Christian Churches to discuss this issue.

The suggested meeting took place, and it was agreed that the following proposals formed a reasonable basis of settlement:

'1 Immediate resumption of work on conditions obtaining on

April 30, 1926, including hours and wages. The settlement, when arrived at, shall be on the basis of a national agreement.

'II A national settlement to be reached within a short defined period, not exceeding four months. In order to carry through Clause I, financial assistance to be granted by the Government during the defined period, under a scheme to be drawn up by the Commissioners who prepared the Report. The Commissioners shall be reappointed for this and for the other purposes mentioned in the following clauses.

'III The terms of the reorganization scheme and the reference to wages in the Report to be worked out in detail by the Commissioners, and the results to be incorporated by them, as far as may be necessary, in a Parliamentary Bill or Bills.

'IV Those parts of the reorganization scheme capable of early application to be put into operation at the earliest moment practicable.

'V The Government to give assurance that those parts of the Report which require legislative sanction shall be placed in the Statute Book at the earliest possible moment.

'VI At the end of the defined period, if disagreements should still exist, a Joint Board, consisting of representatives of both parties, shall appoint an independent chairman, whose award in settlement of these disagreements shall be accepted by both parties.'[3]

The MFGB Executive Committee agreed at its meeting on July 15th, to recommend these proposals to a delegate conference which was to be convened for July 30th. By the time the Conference met, the 'Bishop's proposals' had already been rejected by the Government. Indeed, the Government had now taken the step of amending the Coal Mines Acts so as to permit men to work underground for eight hours instead of seven, so that they were now a very long way indeed from the Samuel Commission's proposals.

However, the importance of the 'Bishop's Proposals' for us lies in the reception given to the actions of the Executive by the various districts. The Executive Committee maintained that they had been tactically correct in co-operating with the Christian Churches in formulating a sensible compromise: the Government had been forced to give a definite decision on the proposals, and in rejecting them had strengthened the position of the miners. As A. J. Cook put it:

'As you know, the proposals were turned down by the Government, and as a result we feel that the intentions of the Committee have been realized and that we have obtained a fresh acquisition of moral support throughout the country. I believe that in the eyes of the public by this action we have been able to put the Government entirely in the wrong.'

However, Durham and South Wales criticized the actions of the Executive. In particular, they were disturbed at that proposal (Clause 6 of the 'Bishop's Memorandum') which provided for a final ruling to be given by an independent chairman; and they considered that before agreeing to such a fundamental change of policy, the Executive should have consulted a national conference. Any suggestion of compulsory arbitration had always aroused vociferous opposition in the Federation. On this occasion Horner put the case against it in these terms:

'Industries where there is a known amount of division can perhaps afford the luxury of an Independent Chairman, but when we are dealing with an industry which is utterly incapable of providing except by outside assistance a decent standard of living for the workmen, and when we know an Independent Chairman is bound to be determined in any judgment he makes by the prevailing facts of the industry on any particular point, then our case will not stand an Independent Chairman who is bound to decide upon the economic facts. . . . I say that the first thing we are bound to have in the circumstances is, perhaps not in Yorkshire but in South Wales, in Durham, a considerable reduction in wages, and we will not be in a position to defend ourselves.'

A. J. Cook for the Executive, agreed with Horner's view of the economic state of the industry, but he drew quite different conclusions. He pointed out that the men and their families were suffering greatly, and that the Federation's bargaining position was weak. There was no hope of the industry being nationalized at that stage, and so long as the pits remained under separate ownership then unless they were prepared to see one-third of the collieries close down (including Northumberland, Durham, South Wales and a part of Scotland) they had to look to the Government for a subsidy. Further, the seven hour day fixed by statute would never be won back until the political complexion of the House was changed, but under the 'Bishop's proposals', a seven-hour day would in fact, be operated by agreement. He went on:

'Unless we can get some means of life in our people then our people will say is this leadership? Somebody has asked, is it leadership to sit still and drift, drift to disaster. That is not leadership.'

He asked the conference

'to face the facts, and come out of the struggle not demoralized, but to retain confidence in each other and not tell everybody that Labour is dead.'

During the course of his speech Cook attacked Spencer who had '. . . said in the House and said publicly in the country that we made a fatal mistake in not accepting the report (of the Samuel Commission). He made a mistake or the Federation made a mistake because he was here. He must be subject to the decision of the Federation just as well as I am'.

In reply Spencer said 'It is true, as Mr Cook says, that . . . I have been facing that report, and always interpreted it to mean a reduction in wages in certain districts.' However, he characterized the speech of the General Secretary as '. . . one of the most remarkable speeches coming from him that this Conference had listened to. It may have taken . . . a very great deal of courage to say at this moment . . . that we must face the facts. I thoroughly endorse that statement. It may have been better, but I am not going to say anything about it, to have said "face the facts", weeks ago.'

Mr Spencer went on to demonstrate the illogicality of the position taken up by Arthur Horner who agreed that an independent chairman—guided by the economic position of the industry—would be bound to reduce wages in the high-cost areas; and yet who refused to accept the Samuel Commission's recommendations as a basis for negotiation. Spencer asked 'where is he going to turn for the means of bridging the difficulty? Is he simply going to turn to the men in South Wales with a policy of despair and continue the struggle? . . . I submit to him and I submit to anybody who thinks with him that they have no positive policy to lay before the Conference this afternoon'.

Spencer referred to the loyalty the men had shown to A. J. Cook —'the most remarkable demonstration of loyalty to a faith, to an individual, that ever has been demonstrated since O'Connor's days in this country'—and he suggested that it would be wrong to reward that loyalty by tearing down any reasonable basis for a settlement of the dispute. He also referred to the virtual impossibility of fixing a national minimum percentage which would be low enough to enable the poorer districts to pay their way and at the same time high enough to satisfy the men in the more profitable districts.

Mr J. Williams of the left wing Forest of Dean agreed with Spencer that Mr Cook's statement was a 'remarkable one'. He continued: 'I wish to submit that when Mr Varley made his proposition with as much courage to negotiate a 10 per cent reduction in wages he did no more than the Secretary has done this afternoon. . . . The least that can be said of Mr Cook is, it is difficult to compare him with what he has said hitherto.' Another delegate, Mr Pearson of Durham also regretted Mr Cook's 'somersault' and thought that the Federation should 'stand solid and firm'. Like Mr Pearson, A. Bevan of South Wales deplored the 'defeatist' speeches of A. J. Cook and George Spencer. He thought that 'It would be better for our men to be defeated as a consequence of their own physical exhaustion than it would be to be defeated as a consequence of any moves we are taking,' and that the policy of Mr Spencer would involve 'conceding point after point to the coalowners', and land them in a position which they could have had on May 1st without coming out at all.

At the conclusion of the debate, it was decided on the motion of Mr W. Carter (Nottinghamshire) seconded by Mr R. Smillie (Scotland) that the 'Bishop's Proposals' should be submitted to districts with a recommendation for their acceptance. South Wales, Lancashire, Durham, Cumberland and the Forest of Dean voted against the recommendation.

The Government and the owners having already rejected the proposals, the issue now was merely one of confidence in the Executive Committee. When the district vote was counted it was found that Durham had come out in support of the Executive, its traditional moderation having triumphed over self-interest, whilst Herbert Smith's home county, Yorkshire, now opposed it, self-interest in this case having been overcome by the county's traditional pugnacity. Two districts, Cleveland and Scotland, did not take a vote so that the Executive Committee was defeated on this issue by 367,750 votes (from five districts) to 333,036 (from thirteen districts). The significance of this vote cannot be ignored. A. J. Cook had met with a rebuff following his advice to 'face the facts'. The districts had decided, if their decision meant anything at all, to 'fight to a finish'. However, it was becoming increasingly clear that this intransigent attitude was being foisted on the Federation by a few large districts in the left-wing strongholds, Lancashire (with 75,000 votes) and South Wales (with 129,150 votes); and the home of plain John Bull, Yorkshire (with 150,000 votes).

In a situation of this kind it was inevitable that demoralization should spread amongst the rank-and-file miners in counties like Nottingham where relatively favourable terms could be arranged without difficulty. The knowledge that a powerful section of the leadership of the NMA believed that an early return to work could, and should, be arranged must also have played its part in forming the climate of opinion in the county as Summer lengthened into Autumn.

In the early days of the dispute, and particularly during the General Strike, the men had been buoyed up by the belief that they were partners in a righteous cause. There was a measure of excitement, with constant meetings and demonstrations, queues for strike pay, picketing and clashes with police. But it is difficult for a hungry man to feel excited, and many men went hungry after the first few weeks.

On May 31st the *Nottingham Journal* opened a relief fund for miners' dependants. With the money so raised, meals were provided mainly for miners' children: sandwiches, the inevitable soup, rice pudding-and-prunes, and similar delicacies were served. At Hucknall the relief was organized by officials of the Urban District Council, and a very efficient job they made of it, too. Elsewhere meals were served in school rooms, chapels, and even private houses of well-wishers (e.g. Mrs Richards, 4 Liddington Street, Basford and Mr Thompson, Ingram Road, Bulwell). Mr J. D. Marsden, the owner of multiple grocery shops, provided 200 gallons of milk a week for young children and nursing mothers, and many other tradesmen gave similar help though on a smaller scale.

Relief of this kind was a help but it did no more than touch the edge of the problem. Indeed, if anything, soup kitchens tend to increase demoralization rather than to add to it. For a man who is proud of his role as his family's bread winner, the thought of his children queuing up (like so many paupers) for soup or rice pudding and prunes must be galling to say the least.

Had there been anything to do, things might not have looked so bad, but for many men there was very little to occupy their time. Those with large gardens were the most fortunate. The summer, happily, was a fine one. But for those without gardens there was nothing much to do except to talk, to play endless desultory games of cards, to pick 'winners' with no hope of raising any money to put on them, to hang about the 'pub' in the hope of getting a free drink, to smoke the odd Woodbine with a guilty feeling that perhaps the money would have been better spent on bread for the children. And always the idle pit wheels stood there to mock.

Everywhere small groups of men stood discussing endlessly the rights and wrongs of the dispute. If they stood too long in a public place they would be moved on by the police.

Sometimes interest would be quickened by a mass meeting addressed by one of the local leaders or A. J. Cook. But the effect of such meetings soon wore off, and excitement gave place once more to apathy.

One observer records this impression:

'Everywhere was the same grim picture. A sad stillness; an appalling quiet at the colliery head, where winding shafts look idly down at little groups of men and women. No smoke; no song; no laughter. Only the occasional rattle of a collector's box or the weird sounds of a quaintly attired jazz band parading the streets (as at Hucknall) for the children's sake.'[4]

3. *The Drift Back to Work*

At the National Conference held on July 30th, William Carter reported that apart from safety men, there were only 500 men in Nottinghamshire at work underground. (Later he amended this estimate to 700.) These men were working at two new pits (Blidworth and Clipstone) which were just being opened out. At Blidworth all the men were working; and at Clipstone, most of the men were at work. Outcropping was described as 'a serious matter . . . which is very menacing and growing and putting a severe test on the men'.

At the next conference, held on August 16th, the position as to men at work below ground was much the same. There was however, a steep increase in the number of men working outcrops. Carter put the figure at 1,700 which was probably on the low side. Near one pit, which normally employed 450 men, some 400 men were outcropping, and this was said Carter, '. . . a very serious difficulty for us, because men are saying that if these men are allowed to continue working outcrop coal they will go and offer their services at the pit.' Throughout the exposed coalfield on the Western side of the county—Strelley, Kimberley, Eastwood and so on—outcropping was going on. With these small, shallow pits history was repeating itself. Some pits were run by teams of 'free colliers' whilst in other cases one man had the contract for one or a group of pits and acted for all the world like a 'big' butty of the eighteenth century. One of the most successful men of this type was Joseph Birkin who ran several pits in the Eastwood district, and made a very good thing out of them despite the instructions of Council—of which he was a member—that outcrop production was to be stopped. In view of the actions of some of its own members it was rather futile to ask, as Linby did, that the transport unions should be requested to prevent the transport of coal during the dispute.

After reporting on the number of men outcropping, Carter went on to say that 'We cannot close our eyes to the fact there is a change in the mentality of the men, and they are asking if something is not going to be done to bring the struggle to a satisfactory termination'. He explained further that the Nottinghamshire Council had, by a majority vote, decided that the county's delegates should be given '. . . every possible latitude to give almost every possible license to the Executive Committee to do everything they possibly can in the negotiations with a view to getting the maximum offer from the powers that be, and that when they think the time has arrived when they can get no further concessions, these proposals to be submitted to the men for acceptance or rejection'.

Certainly by this time, there was a very strong feeling in the county that no useful purpose could be served by prolonging the struggle. Sensing this, the owners wrote to the NMA on August 6th asking for a joint meeting to discuss terms on which employment could be resumed. Council at its meeting on August 12th referred this matter to branches. At its next meeting on August 21st, Council resolved 'that after hearing reports of delegates from all collieries, we urge all members to adhere loyally to the national policy and decline returning to work as terms offered by Employers are altogether inadequate'. Nevertheless, at a further meeting held on August 28th, Mr Varley was asked to 'ascertain whether the Federation Committee at its meeting on Monday next, will permit us to meet our owners locally re terms of a settlement and officials be instructed to call a special council meeting if they think that it is advisable'.

Mr Varley knew, from his experience at the national conference of August 16th, what sort of reception he could expect. On that occasion the fracas had started with a speech from the fiery agent of the Leicestershire Miners, J. W. Smith, who accused the leaders of Nottinghamshire of disloyalty to the Federation: He hinted that Nottinghamshire was contemplating a district settlement, and he invited the people concerned to come out into the open instead of employing 'back-door' methods. Then the chairman, Herbert Smith joined in. He said:

'Then we have Mr Varley. It is not the first time Mr Varley has put us in this position, and I am just about tired of it.'

'MR F. VARLEY, MP (Nottinghamshire) Fishing is not quite good enough. You say it is not the first time you put us in this position —in what position?'

'THE CHAIRMAN: I shall say what I am going to say in my own way. You can take your exception to it in your own way. I am putting the position now. Mr Varley came before the Executive, which he has a right to do, and put the position; suggested a reduction in wages, but does not go to the Nottinghamshire Council when the Executive turned him down, but he goes to the public and gives his opinion, and writes an article in the Press quite contrary to what the Executive had said to him. Last week I put the question twice. I had more than one in mind. There were Frank Hall and Varley. My question was: Is it a fact that some people are talking about, or considering about district agreements? It did not take hold. I put it a second time. It did not carry again. When I got a paper we had Varley recommending the men to go and meet the owners without committing themselves. I cannot see how you can meet the owners without committing yourselves. . . . I do not believe Varley has played fair in this matter.'

In reply, Mr Varley said that Nottinghamshire had not received any offer from the owners since April 30th, although they had been asked to meet the owners on three occasions. He thought that they were bound to do something in Nottinghamshire or the union would disintegrate. He went on:

'It is far better to keep hold of this thing at the first stage, get to know what the offer is, and then come to the conference and report. And if the Midlands can be made to meet the position of a national settlement, that is what is intended by the people behind the move. I have never stood behind slogans right from the beginning. I have said we have got to compromise on wages. I have stood up for hours and a national agreement, and I am in no apologetic mood, and I have made my position clear to my men.'

Next, Arthur Horner added his word of condemnation: He was opposed to the resumption of negotiations even on a national basis. Instead, he suggested that the Federation should stop safety men from raising coal; tackle the outcrop problem, concentrate its speakers on those areas where many men had returned to work, and prevail upon the transport workers to stop the transport of imported coal. Other speakers from South Wales also opposed any resumption of negotiations, whilst Mr Peter Lee of Durham advised Conference to 'watch the Midlands, that is the weak spot, not because they are worse men, not because they are worse organized, but because they are separate districts and have been hard hit'.

Eventually Conference accepted by 428 votes to 360 (the minority consisting of Forest of Dean, Lancashire, South Wales and York-shire) a motion submitted by George Spencer on behalf of Notting-hamshire which reads:

'That this conference instructs the Executive to open up nego-tiations either with the coalowners or the Government or both, for the purpose of securing the maximum terms from them, and to submit to a further conference the results of their efforts, but with-out settling anything themselves.'

George Spencer then suggested 'that the power be given for un-restricted negotiations'. Presumably he meant that the Federation representatives should be prepared to negotiate on hours, on wages and on the type of agreement (e.g. they should not insist on a national minimum percentage). This suggestion met with an extremely un-favourable response however, and was not pursued.

At the conclusion of the debate, an opportunity was given for J. W. Smith of Leicester to criticize Nottingham's Frank Varley, and for the latter to reply. Unfortunately, Mr Smith tended to ramble a little and it is therefore difficult at times to understand exactly what he is saying. However, the general nature of his accusation may be gathered from the extract from the verbatim report given below:

'Mr J. W. Smith (Leicester): It was on the 30th of last month that this discussion [with Varley] took place, and I think I referred to the proposals in the Conference on the 30th of last month. My men were instructed to sign on the following Wednesday, August 4th. It left me only two days, three including Sunday, to deal with these proposals and keep the men solid against a break-away. I want you to realize my difficulty, when we have leaders in some districts who have said amongst other things that Herbert Smith had crawled on his belly, and the sooner there were district settlements the better. Mr Varley's name has been used, difficulties have been created with the rank-and-file and, let me say this, that the rank-and-file have been solid enough up to now. The question this conference must answer is this: is Nottinghamshire to be tolerated in the move they have made which makes for a break in the Federation? Are they to be tolerated in the face of the national situation, at the same time of claiming the right to decide a question which is not its own question. If the Conference says No to it, then I shall take it back to my Council.'

'Mr Frank Varley, MP (Nottinghamshire): It seems now that it may be very well to explain the manner in which the question arose. Yesterday it was that I had perverted the pristine simplicity of Mr Smith to use him for some subterranean development. I am not going to pursue it further except to say that I was in the "White Hart" when he came in and referred to the proposals made to his district which formed the basis of discussion. . . . These proposals made Mr Smith fear that the very excellence of the offer would weaken his men. I am not concerned as to what the newspaper says. We had two meetings on the following Friday and ostensibly they had seen the proposals, but Mr Smith had discovered that the offer was not so good as he thought, and had taken fresh courage. I have never communicated with Mr Smith, who seems so alarmed about getting me as an adherent in some kind of subterranean development. It is characteristic, because when he disagreed with his Association in Nottinghamshire he came and broke the office windows. I will ignore that. Now, as to whether Nottinghamshire is to be allowed to do what she proposes to do. I want to suggest he has got his proposals considered by his own

Council, but we have no proposals. We have heard nothing since the 30th April as to what the intention of our owners is. We are inspired by the fact that we have got to do something. We believe that our motives are sincere and we have got to keep hold of it, otherwise they will take it in spite of our efforts.'

G. A. Spencer explained Varley's final sentence.

'We have had for three weeks,' he said, 'people attending conferences who live in the district, active Trade Union members; for three weeks there have been unofficial deputations of men meeting the owners of that county for the purpose of arriving at an agreement.'

Consequently, when the owners had requested a meeting with the Association, Council had submitted the request to Branches for a decision in the belief that, if they were to reject the owners' request out of hand, the reins of power would pass to the unofficial negotiators. Clearly, there was already in existence an embryo breakaway, although as yet it lacked strong leadership. Spencer and Varley saw the danger of allowing the initiative to slip into the hands of the breakaway movement. However, Spencer went on to say that, since the MFGB Conference had accepted the Nottinghamshire resolution calling for national negotiations to be reopened, he was sure that the NMA would not now meet the coalowners locally. He continued:

'If the decision had gone the other way and that we were going to do nothing at all, I don't know what the position would have been. We should probably have found the influence of our county too strong, no matter what our own minds might be.' He assured Conference that he was '. . . as desirous of having a national agreement as any man in this conference.'

That was on August 16th. Three days later the MFGB Executive met the Mining Association. At this meeting the owners made it quite clear that they were not prepared to negotiate nationally. Reduced pay and increased hours were necessary, but the details would have to be worked out in the districts. The Chairman of the Mining Association, Mr Evan Williams, indicated that the district owners' Associations were no longer willing to allow the national body to take any part in wages issues and he saw no point therefore, in arranging any further meetings with the Miners' Federation at national level.

A week later, on August 26th, the officers of the Federation met the Chancellor of the Exchequer (Mr Churchill) and other Government representatives; but this meeting was also unfruitful. The

Nottinghamshire resolution on reopening negotiations nationally had therefore proved to be abortive; hence the decision taken by Council on August 28th to seek the Federation's permission to negotiate locally.

At the national Executive meeting held on August 31st, Frank Varley explained the position in Nottinghamshire at some length. Doubtless, he underlined the warning given to the August conference that failure to exercise positive leadership on the part of the Association would result in a transfer of leadership to those who were prepared to negotiate an unofficial settlement. However, Herbert Smith explained:

> 'that he could not put the question to the meeting as it was directly contrary to the Federation policy and, therefore, out of order; the policy could not be changed by the Committee, but only by a Conference, and therefore it would not be in order for the Committee to consider the request.'

Two days later, a national conference assembled in the Kingsway Hall, London, heard William Carter report that there were 7,000 to 8,000 men back at work in Nottinghamshire, ten times as many as there had been a fortnight earlier. In addition, there were about 1,500 men on outcrop workings. Sixty-five per cent of the men at the Bolsover pits were now working—a matter of between six and seven thousand men. The delegates of Clipstone and Crown Farm (both of which are Bolsover pits) H. Willett and William Holland were among those referred to by George Spencer at the previous conference, who had been discussing possible terms for a return to work with the owners. At this time, as William Carter made clear, George Spencer in common with the other Permanent Officials, continued to address meetings of the men asking them to stay loyal to the Federation. The Association had also had the help of A. J. Cook whose oratory had undoubtedly stiffened the resolution of many waverers.

Nevertheless, many men had gone back to the pits. Carter explained that 'some three or four of the collieries had imported police, and charges are being made of a wholesale character; men brought to the courts on the most trivial offences; there is almost a state of terror in these particular districts'. The Bolsover men had gone back on quite favourable terms. They were now working $7\frac{1}{2}$ hours a day instead of seven (and in consequence the pieceworker's addition for working less than eight hours a day was now reduced from roughly 14 per cent to 7 per cent). The minimum percentage on basis rates was to remain at 46·67 for seven months, after which there was to be a new minimum of 36 per cent. Similarly, the subsistence rate for stallmen was to be reduced from 8s 9d to 8s 3d, and the proportion of the net proceeds going to wages was to be reduced from 87 to 85 per cent, the proportion going to profit being increased from 13 to 15 per cent. Carter warned the Conference that:

'. . . the number going back is increasing and not decreasing. They are looking forward to some change, and we hope that if the change takes place, that (it) should take place from headquarters. We cannot afford to ignore this thing, more men going back, hundreds and thousands; it is going to be a most difficult matter to collect the scattered forces unless something is done, unless we have some discipline.'

Earlier, Mr George Annable of the Enginemen's union had reported to Conference that many of his members in Nottinghamshire—including the Bolsover men—had gone back at their pre-stoppage wages. Speaking for Derbyshire, Mr J. Spencer reported that some 3,400 men were back at work in his county, the Bolsover pits again being the 'black spots'. The position in Nottinghamshire had however, created difficulties particularly in the Mansfield district where both County Associations had members.

'We are up against a very serious position in helping our men having regard to what is happening in the Mansfield and in some parts of the Nottinghamshire coalfield',

said Mr Spencer. He went on with a veiled reference to Frank Varley:

'I do feel this: if we cannot all be brilliant and I do not profess to be brilliant, we ought to be honest to the policy of this Miners' Federation and when we adopt a policy, we ought to carry that policy out until the Federation Executive proposes some other scheme, and that is the position of Derbyshire.'

For the moment, whilst both Varley and Spencer were known to favour an early return to work on the most favourable terms obtainable, there is not a shred of evidence to suggest that either of them had helped in arranging a settlement. Indeed, all the evidence points the other way. At the beginning of September, the permanent officials of the NMA were still publicly supporting the policy of the Federation to the extent of urging the men to stay out. At the same time, they did advocate compromise solutions, believing that the battle was now as good as lost. They felt that leadership was slipping from their grasp, and that the only way in which they could recover control from the secessionist delegates was to 'swim with the tide'. Let us repeat, whilst they were prepared to urge the men to stay out in loyalty to the Federation, they were not prepared to hide their desire

M

for a compromise solution. As we have seen, Varley had suggested one way out, involving a wage-cut, as early as May 28th (in his *New Leader* article) and now Spencer in a speech in the House of Commons on August 30th, pointed the way to another possible solution involving the abandonment of the claim for a national minimum wage. The President of the MFGB, Herbert Smith, at the Conference held on September 2nd, took exception to Spencer's speech, and in particular to the place in which it was made: Spencer 'ought to have made this speech in this Conference or in Notting-hamshire Council before making it in the House of Commons'. The particular part of the speech to which Smith took objection is as follows:

'We have not had a national agreement since the war. Anyone who says we have had a national agreement is not really repre-senting the true facts of the case. All we have had is a national formula locally applied, giving various results. How can you have a national agreement which gives you in one district 50 per cent more wages than in another? That cannot be a national agreement and you have not had a national agreement. You have had a national formula. But you have had this, which has given it the look of a national agreement. You had a national minimum wage in that agreement. You cannot maintain a national minimum wage unless you have a protected subsidy. I cannot see the end of the subsidy if you are going to pay the present minimum wage, but you can have modifications of the minimum, as applicable to the economic character of the districts and at the same time preserve your national agreement, and your national formula with regard to the proportions that we are to have. The real virtue of a national agreement to us is that it strengthens our bargaining power. It is a simultaneous attempt to impose conditions on each district. Under the old conditions, you would have one area trying to get an improvement of one kind and another area another, and the bargaining power of several districts is not commensurate with the bargaining power of the whole Federation. I see no reason why you could not have a national agreement, but accommodation must be found for varying minima so far as other districts are concerned.'

Later, Arthur Horner of South Wales wanted to know: 'what is going to happen with respect to the Nottinghamshire district. Are we going to stop this district manoeuvring?' Horner went on to say that Nottinghamshire had given 'official sanction' to breakaways, a statement to which Mr Carter took exception since he and his colleagues were 'not only working night and day but every day in the week with a view to trying to keep the men solid'. George

Spencer explained that there had been a motion before Council to enter into direct negotiations with the district owners' association, but that this suggestion had been turned down in favour of referring the issue to the national conference. He ended:

'There is not one of us want local settlements—not one of us— we want a national settlement. We may disagree upon points of detail, but we want a national agreement in Nottinghamshire as strongly as anybody else, and will do everything possible in the next fortnight to keep the Association intact.'

There seems to be no room for doubt that the Nottinghamshire officials were doing their best at this stage to avert a breakaway, but they were struggling against the tide. A rank-and-file delegate from the county warned Conference that there was a likelihood of at least another 10,000 men going back during the following week (i.e. the first week in September). He went on to say that whilst the Leen Valley was still solid, the other two parts of the county (the Erewash Valley and the Mansfield district) were weak.

On September 3rd, the day following the Conference, the Executive Committee of the MFGB together with Mr Ramsay Macdonald met the Chancellor of the Exchequer. As a result of the discussion at this meeting, the following letter was sent to the Government:

'Miners Federation of Great Britain.

'Rt Hon. Winston Churchill, MP,
   10, Downing Street,
      London, S.W.1

'Dear Sir,
   'I beg to inform you that the Executive Committee and the Special Delegate Conference of the Miners' Federation, having again carefully considered the present deadlock in the mining dispute, have resolved to ask you to convene and attend a Con- ference of the Mining Association and the Federation. We are prepared to enter into negotiations for a new national agreement with a view to a reduction in labour costs to meet the immediate necessities of the mining industry.

                              'Yours faithfully,
                                (Signed) A. J. Cook, Secretary.'

Unfortunately, this significant alteration in the Federation's stand was a wasted gesture. Mr Churchill, who was probably responsible for the form of words used in the Federation's letter ('a reduction in labour costs to meet the immediate necessities of the mining industry') met the Mining Association on September 6th, only to be faced with a blunt refusal to reopen negotiations on a national footing. The coalowners' leader, Mr Evan Williams said that national agreements led to political interference in the industry, and the owners were not prepared to countenance this any longer. He pointed out that the national agreement had brought about 'a stoppage even where there is no dispute as regards wages between the owners and the men', and he instanced the counties of Nottingham, Derby and Leicester. He went on to say:

'it is a political issue pure and simple, and the moment you have set up a national agreement with a national board you bring every question that is relevant to that board forward as a political issue, with debates in the House of Commons, and you get the Government involved and you put the industry as a whole on a different plane from every other industry, and such accusations as have been made that there are troubles in the coal industry and that they are more numerous and more bitter than in any other industry of the country are due solely and entirely to the fact that they are dealt with on a national basis and not on a district basis.'

He also insisted that it had been understood between the Government and the Owners that the latter would offer the best possible terms for a return to work in each district separately, thus inducing 'breakaways' in the more prosperous counties. Faced with so intransigent an attitude, there was nothing Mr Churchill could do to help.

And so the drift back to work continued. By the time the next national conference was held (September 29th) some 16,000 to 17,000 men had gone back to work in Nottinghamshire on the union's own computation, and this was probably a conservative estimate. In other parts of the country, there were signs of weakening, particularly in the Midlands. According to the figures supplied by the district associations, there were in all 81,178 men back at work compared with 36,785 on September 2nd. The bulk of these men (60,600) belonged to Nottinghamshire, Derbyshire and the Midland Federation.

The conference discussed one possible way out of the impasse— to accept an offer made by the Government on September 17th. The Government proposed that the men should return to work on district agreements, but that following a general resumption of work, a National Arbitration Tribunal would be established by statute. This Tribunal would be empowered to review any matters to do with wages and conditions of employment submitted to it from any district where more than seven hours a day was being worked. George Spencer moved:

'That this Conference instructs the Executive Committee to

immediately consider the proposals of the Government with a view to making a recommendation to the districts based on these proposals and report to us tomorrow morning.'

In moving the motion he said:

'. . . we have either to accept the Government's proposals or fight on. Now, if we are going to fight on, does anyone think for one moment we are going to succeed in achieving our original object? Is there any man in this Conference who thinks that? I think we had better face realities, and what are we going to gain by prolonging this struggle for another fortnight or three weeks? As a matter of fact we shall be in a worse position at the end of three weeks than we are today because of this fact, you must take account of the dribbling back. . . . You may say "Be loyal to the Federation". We have endeavoured against strong pressure to be loyal to the Federation. . . . Whilst we have a duty to the Federation, we have a duty to the men we immediately represent.

'. . . Some people have dribbled back to work, and I would rather we assume now a joint responsibility and say to the men, having done all we possibly can, we cannot succeed in attaining our end, and advise them to take these proposals as being the best possible way of achieving something and saving something on their behalf.'

In reply, Herbert Smith made it clear that if Spencer's motion were carried, the Executive Committee would bring back a recommendation to reject absolutely the Government's proposals. This was underlined by the Treasurer, Mr W. P. Richardson (Durham), who said '. . . there is not a single person on that Committee ever given the slightest hint in favour of the Government's proposals, and therefore, to send the thing back is simply dilly-dallying all the time, and would serve no useful object'; whilst Mr Pearson, also of Durham, thought that Spencer was insulting the Committee in moving his resolution. Subsequent speakers also spoke against the motion.

On the following day, September 30th, it was stated that various members of the Executive Committee would proceed to Nottinghamshire on the conclusion of the Conference to engage in propaganda work. Of this suggestion Spencer said:

'We are all very grateful for the assistance which has been rendered in Nottinghamshire and Derbyshire, but if anyone is coming they must have a new story—a new tale to what some have been telling, because so far as we have gone up to now we have had six hours, £1 a day and Sankey, and if anybody else comes there must be something better than that to put before the men.'

He went on to complain of speakers coming into the district without the consent of the county officials. Dealing with the suggestion that the NMA was about to conclude a district agreement with the owners he said:

'. . . It will all depend on what our Council meeting says. If we find tomorrow there has been such a body of men returned to work, that there is more than a majority of the men in the county returned to work, perhaps by next Thursday they have all returned to work, *they may say we have got to make an Agreement to preserve our Association.* . . . We shall be guided tomorrow morning by the circumstances, and if we find there is still a possibility of holding the great majority of the men in the county, we shall endeavour still to be loyal to the Federation, but are we going to sit still and have the whole of the men at work before we move? If we can do anything to stem the tide, or anybody else can, we shall be very thankful for it so long as they come with a mandate, but we cannot tolerate men coming into the district simply to make Spencer and Varley a text for their speeches. Some of the men have gone to work as a result of the speeches which have been made, and if the same speeches are made in the next few days, it will have the effect of sending some more men back to work.'

Supporting Spencer, Varley said that the NMA had had to 'foot a bill to the extent of £115 in expenses for people coming into our county since August 22nd'. He went on:

'Whatever we have lacked, it has not been oratory. They have been at it from early morning to dewy eve, and right into the hours of darkness. We have had a demonstration like the Duke of York —something attempted and nothing done, and did not earn a night's repose.'

Mr Varley went on to say that the Mansfield and District Committee—the semi-official left-wing committee to which we have referred before—had been given a free hand, with the result that there had been too many speakers of one sort (he instanced A. J. Cook and John Symes) and he concluded:

'We have feared this district right from the commencement that they would break us up there.'

The temper of the Mansfield district was to some extent the result of an influx of men from South Wales which took place in 1923. The Kirkby pits had a very large proportion of Welshmen, and there were many also at the other pits around Mansfield.

William Carter explained to Conference that resolutions advocating a district settlement had been coming in from Branches, but that the officials had done their best to hold the men pending a move at national level. Council would be advised not to entertain a district settlement, but if they decided to 'go over the heads' of the officers, that would be another matter. Carter looked forward to the assistance of the three national speakers who were coming into the district—Mr T. Cape of Cumberland, Mr Joseph Jones of Yorkshire and Mr Smith of the Enginemen and Firemen's Union.

On the day following the national conference (October 1st) Council met under the chairmanship of Mr Ben Smith the Vice-President. This was to be Mr Smith's last meeting since, shortly afterwards, he was to return to work. Many other delegates had already gone back and had therefore been excluded from Council.

Council had before it a further small crop of branch appeals calling for the resignation of the officials, but its deliberations now wore an air of unreality. The one positive decision taken—a recommendation that the Government's latest proposals should be accepted—was turned down by the branches, twenty-six voting against to four in favour. This result is easily understood when one appreciates that those members who were back at work, or were on the point of going back were no longer, generally speaking, taking any part in the Association's activities so that the decisions taken tended to reflect the wishes of the militant minority.

For Nottinghamshire, the strike was now drawing to a close. The delegates present at the Council meeting of October 1st reckoned the number of men at work in the county to be in the region of 32,000 to 33,000: roughly double the figures given by Mr Carter to the national conference held the day previously. Of the Association's 34,000 members, roughly 70 per cent were back at work. The dilemma which now faced the district officials was a very real one: were they to exhort the odd 30 per cent to stay out knowing perfectly well that in so doing they would run the risk of becoming permanently unemployed, or were they to advise these men to safeguard their livelihoods by returning to work whilst their old jobs were still open for them? Would it be right to ask that men who had stayed firm throughout the dispute should commit economic suicide by remaining loyal to a lost cause? Would it not be better to arrange a return to work which would provide for men to go back to the jobs they had had before the dispute?

Such questions as these were in the air in the first week of October 1926.

4. *The 'Breakaway'*
On Wednesday, October 6th, the *Nottingham Evening Post* reported

that a meeting of the Digby and New London men held on the previous evening had decided unanimously on an immediate return to work. There was nothing particularly startling about this decision —other lodges had done the same without so much publicity. The significant feature of the Hill Top meeting was the presence of Mr G. A. Spencer who was:

'. . . authorized to seek an interview with the management. This Mr Spencer did this morning, with satisfactory results, and the men are being notified to present themselves for work tomorrow morning. The men, we understand, will return to work on the same conditions as those already at work.'

It appears that on Monday, October 4th, about 490 of the 600 men normally employed at Digby and New London were back at work. A meeting of those still out was held that day, and it was agreed that they should remain loyal to the Federation's policy. However, during the course of the day more and more men signed on, and those who were left panicked. On the Tuesday morning therefore, the Lodge Committee was instructed to seek an interview with the management in order to arrange for a general resumption. The deputation requested the manager to allow all the men who had been on the books before the strike to go back to work at their old jobs, but this the manager refused to do. Instead, he said that every man would be allowed to go back to his old job, provided that no one else had it already. Obviously by this time all the best jobs would be filled by those who had 'blacklegged', and the chances were that some of those who were still out would not be taken back at all.

The Lodge Committee thereupon resolved to prolong the dispute, and they tried to bring all the men out again. However, only thirteen of the 400 who were working would agree to come out; and of the 200 who had stayed loyal, only seventy would promise to remain so. It was in this situation that Mr Spencer addressed the Hill Top meeting. According to the Branch Delegate, Mr C. A. Pugh, 'Mr Spencer deplored the action they had taken'. However, he agreed to see the manager on the day following to see whether he would be prepared to take all the men back. This the manager now agreed to do, and practically all the men therefore signed on.

Spencer was then approached by men from other pits, and during the course of the day he arranged with the managers of seven other pits for all the men to be allowed to resume work.

That was on October 6th. On the following day the matter was raised at a national conference of the MFGB. Mr A. J. Cook asked whether it was true that some of the Nottinghamshire Lodges had met and arranged local terms of settlement. If this was so, then they had to face a situation in which part of the Federation was 'officially blacklegging'. The President wanted to know whether the NMA condoned the action which had been taken in which case he suggested that Nottinghamshire should seriously consider whether they should take any further part in the national negotiations. In reply, Frank Varley admitted that some local settlements had been made, and said that the NMA had no option but to condone the actions of those concerned. Whilst it was true that Council had a few days previously rejected a proposal that a district agreement should be negotiated, yet half the delegates left the meeting with the firm intention of making arrangements with their employers at the risk of being 'disfranchised' by the NMA. Varley went on to state frankly that his county was now out of the strike, many pits being back at full strength. However, both he and Carter were worried at the suggestion that Nottinghamshire should be turned out of the Conference. Carter suggested that the NMA should be allowed to call Mr Spencer to account for his action, and this was agreed to. It was also agreed that Spencer should be summoned to attend Conference on the following day.

Meantime, Carter and Mr T. Cape of Cumberland (who had been working in Nottinghamshire on the instructions of the Federation) tried to put the recent events in Nottinghamshire into perspective. Mr Carter explained that many of the loyal trade unionists had been compelled to go back to work 'against their conscientious convictions . . . with broken hearts'. Some pits with a pre-stoppage labour force of 1,000 or 1,200 men were now saying that they had their full complement whilst 200 or 300 men who had stood loyal were faced with the possibility of permanent unemployment. Further, those who were back at work were without union protection, and victimization was taking place. Mr Carter warned the conference that unless the National Executive made some proposals for a settlement, Nottinghamshire would return to the 'chaotic and unenviable state of having pit agreements'.

Mr Cape compared the task of stopping men going back to work, and getting those who were already back at work to come out again, with that of King Canute. He referred to the situation in the Leen Valley proper where four of the five big pits had decided to return to work following meetings held on Sunday, October 3rd. The men at the fifth pit had decided to stay out but found to their chagrin that their places were being taken by men from other districts. These men who were standing loyal were subjected to 'every kind of persecution, intimidation and persuasion'. Some employers had warned the men that unless they returned to work by a stated date, they would receive their insurance cards and indeed, many men had received their cards already. Mr Cape had never seen so many police in any locality before. If two men stood to speak, they were moved on.

Mr Cape also referred to the peculiar state of affairs which had arisen at New Hucknall, where an attempt had been made to supersede the checkweighmen who had refused to start work when their men went back. At a meeting of the members of the Checkweigh Fund, it had been decided that the two men who had been appointed to take the place of the old checkweighmen should not be paid. It was further agreed that the old checkweighmen should be allowed to stay out pending a national settlement, and that they should then be permitted to return. Further, if the management refused to allow the checkweighmen to take up their duties when the time came, then the men would come out on strike again. This was a remarkable demonstration of loyalty. Force of circumstances had driven the men back to work, but this did not prevent them from showing in a practical way their admiration for the stand taken by their local leaders.

On the second day of the conference, Friday, October 8th, Mr G. A. Spencer attended Conference. The sort of reception he was likely to get was indicated by an incident which occurred within a few minutes of the opening of the session. The President, Mr Herbert Smith, asked whether it was true that one of the Nottinghamshire delegates was back at work. William Carter admitted that this was so, and explained that the delegate concerned (Mr W. Buckley of Wollaton) had indicated his intention of returning to work to Council, and had offered to resign. However, his resignation had not been accepted, and he therefore attended Conference with the authority of the NMA.

Mr A. J. Cook said that Nottinghamshire had placed him in an awkward position. He had been asked to go to the Mansfield Area to try to bring the men out again, and he had just received a message from Mansfield that some 800 men had once more downed tools, but that the talk of the district was that they had a blackleg in the Conference of the Miners' Federation. It was moved and seconded that Buckley should be asked to leave the Conference. After Mr Bunfield, Treasurer of the NMA had agreed that the delegate should go, but had suggested in so doing that 'a poor sinner wants sympathy sometimes', the following exchange took place:

'MR BUCKLEY (Nottinghamshire): I think this policy is one of dividing Nottinghamshire. Are we the only district in the Conference?

'MR COOK: I object to a blackleg speaking in this conference.

'A DELEGATE: Let him put his position.

'THE CHAIRMAN: There is no position to put at all. What this Federation decided to do along with Nottinghamshire on the 30th of April, and before, was to resist a reduction in wages, resist an extension of hours, and to resist district agreements. You people have gone and sanctioned these things, and you are one of them. I ask you to leave this Conference.

'MR BUCKLEY: Will you allow me to explain?

'THE CHAIRMAN: No.

'MR BUCKLEY: I can leave the room, but I want to thank the Area Committee of Mansfield—Mr Cook knows.

'MR COOK: I object.

'MR BUCKLEY: I can give an explanation of the whole thing. I will leave the conference, but this is the best way to smash Nottinghamshire up.'

Mr Buckley then left the conference, and the chairman turned his attention to Mr Spencer who was asked for an explanation of his actions at Digby and New London. Mr Spencer said:

'I am not very good at speaking this morning. I am not in the very best of form as I was in bed yesterday when I got your wire, but I have done my best to get myself right so that I could come this morning, and I am here. I will first of all deal with the facts and I am not going to say I am not guilty—I am guilty; let me be straight and give you exactly what the facts are. Mr Varley answered the telephone that morning and said, "You are wanted at Digby". I went to the 'phone, and I was asked: "Will you come and address a meeting tonight." I said, "I will." I went over to the particular pits and addressed the men. A good many of them had gone to work, I think the majority, I don't know. A delegate is here, if I may say so, who belongs to the same pit, who was at the same meeting, and he will be able to vouch for everything I have got to say. I addressed that meeting that night. It was not only the men who had actually gone to work, but a good many other men who had not gone to work but were going the next day, at not only these pits, but other pits in the immediate district. The unfortunate position was this, that the managers in the district were absolutely refusing to sign on men who had been most prominent and most faithful to the Federation, so I was asked that night when they heard what I had to say whether I would, after passing this resolution, go with them. I want the delegate here to say whether I am speaking the truth. I told them it was a very difficult position to put me in but I was not going to play the coward. If they wanted me to do it, I was prepared to sacrifice myself on their behalf. Now the men in that locality, if they had not gone to work, it meant men who have been loyal to the Federation would have been victimized. There were men in that district who were nine and twelve months after 1921 before they got back again to work. So I went. I went to this particular pit. I had a deputation because the manager refused to sign certain men on. I don't want to bring anybody else into this. I want to keep it to myself. The delegate at that particular pit [Mr C. A. Pugh] has been one of the most ardent supporters of the Federation policy, but he has had to take up this position and the reason he had to take his share—and by taking his share and returning to work—was because he saw so far as that locality was concerned it was hopeless, and they had to solve their own particular position. Under the circumstances, I went with them. I don't regret it and I do not plead extenuating circumstances. I believe I did the best day's work in my life for these men, and you can pass your sentence.'

The President, Herbert Smith, in reply accused Spencer of cowardice in going away from the line of policy laid down by the Federation, and said:

'. . . If Mr Spencer does not regret it, I am sorry for him, and I say this Federation ought to deal exactly with Mr Spencer as with anybody else. . . . Whether it is cowardice or not, I would rather be shot in the morning than do what you have done. You are a paid servant of this Federation.'

Mr. A. J. Cook went further. He said:

'I hope this Conference will treat Mr Spencer as they would treat a blackleg. Mr Spencer is a blackleg of the worst order. A conscious blackleg. I want to say here that Nottinghamshire has been more responsible for the present position we are in, and Mr Spencer is more responsible for Nottinghamshire than any other district in the coalfield. . . . I think we ought to ask him to leave the conference and excommunicate him as we would any other blackleg.'

The Digby delegate, Mr C. A. Pugh, tried to put the matter in perspective. He said:

'. . . There is no love between me and Mr Spencer. I have fought him all along. . . . I have been quite as extreme as any of the elements in South Wales, but when men come to me as their representative, as their delegate, and say: For God's sake do something for us . . . or we are going to be thrown on the streets . . .

Mr Spencer came along. We asked the men who were at work if they were prepared to come out again. I think seven or eight got up, and we asked the other men who were standing solid if they were prepared to still stand solid. There was a hesitancy which meant probably the next morning the majority would have been at work. As a last resort the Committee asked Mr Spencer to go on a deputation to see the management to obtain the right for these men to be set on again and not victimized.'

In discussion, Mr Spencer stated that Mr Pugh had moved the resolution to return to work, and that he had signed on. Later, Mr Frank Varley referred to Mr Pugh as a Communist. This raised a protest from Mr Arthur Horner of South Wales who said:

'I want to know what is the object of this continual imputation of the political party to which a man happens to belong, without any regard to that which other people belong. I was never more ashamed in my life. He will be dealt with. I want to know whether the ILP will deal with Mr Spencer, and whether or not other organizations will be mentioned as well as the Communist.'

The motion that Mr Spencer should be asked to leave the conference was carried by 759 votes to four, Cleveland being the only district to vote against. Following the conference, Spencer went to see one of the founders of the NMA, and the Association's printer, Alderman W. Mellors of Hucknall. Mr Ernest Mellors, son of the Alderman, who was in partnership with his father was present at the interview and he describes what happened thus:

'Mr Spencer came into the printer's yard looking worn and ill. He sat on a chair outside the workshop, and completely broke down. He told Alderman Mellors that he considered his career to be at an end: in doing his best to safeguard the livelihoods of his members, he had lost everything he had been fighting for over the years. He was bitter at being forced into a false position by the extremist element in the Federation. Alderman Mellors apparently sympathized with Spencer, and thought that he had had little option but to do as he had done over the Digby affair.'

Spencer had been turned out of the National Conference, but he still had to face the Nottinghamshire Council. The letter calling him to the Council meeting from Mr W. Carter stated, according to Spencer's own account, that Spencer '. . . should be exonerated from all blame' and that what he had done 'had been in the interests of the men themselves'. The President of the Association, Mr Val Coleman, was away in Russia as a delegate of the MFGB whilst the Vice-President and twenty-seven other delegates (out of a total of forty-two) had left the Council because they had gone to work.

Council met on October 16th under the chairmanship of Mr W. Bayliss. According to Spencer himself, Mr Bayliss refused to read Carter's letter which Spencer regarded as a 'vindication', and Council resolved that Spencer should be suspended until Council had been re-constituted, then his case should be considered.

The developing split in the Nottinghamshire Miners' Association was causing grave concern to those who had spent their lives in building up the organization. And there is no doubt that, had the local officials been allowed to decide the issue free from outside pressure an accommodation would have been come to. It was obvious by September that the stoppage could not be successful, and that before long, privation would drive the miners back on the owners' terms. Under these circumstances, had the NMA been prepared to refrain from taking precipitate action against the strike breakers, its unity would not have been permanently impaired. However, pressure was put on the NMA to loyally support the policy of the Federation: and to expel those who were 'blacklegging'.

On September 25th Council resolved that 'all Delegates who have not carried out Federation policy be suspended, and those who have carried it out be allowed to remain'. At the same meeting, the permanent officials were instructed 'to call a meeting of the members of the Clipstone Branch with a view of inducing the Clipstone men to take the step of superseding the present Local Officials with others who will carry out the duties of the organization'.

However, at its next meeting, Council was considering whether to take subscriptions from those members who were back at work. Had this been done, the Association would it is true, have given a certain measure of recognition to the strike breakers; but it should be borne in mind that most of the men concerned had gone back to work after standing loyally by the Federation since May 1st, and doubtless they felt that since the struggle was hopeless anyway they were now justified in considering the equally strong loyalties which family life imposes. Let us say then that, had their loyalty not been put to such a very severe strain, these men would have remained model trade unionists. Under these circumstances, there would have been little harm done by collecting their contributions: the money would without doubt have been welcome to those still out on strike. However, their subscriptions were not collected, and many of them were lost permanently to the Association.

On October 11th, Council adopted the following motion:

'Resolved that the following information and form of ballot paper be issued to members.

TO ALL WHO WERE MEMBERS ON APRIL 30th, 1926.

'It has been found impossible to carry on the business of the Association owing to some Branches being at work, and others standing firm, you are asked to give your decision on the following:

'1. Are you in favour of again ceasing work, standing by the Federation Policy, and disfranchising all who do not act accordingly.

'2. Are you in favour of seeking a District Settlement, which may entail severance from the Federation.'

The issue of this circular brought a protest from the Mansfield Area Committee who said that a vote for a district settlement would 'give Mr Varley and Mr Spencer another lease of life to continue their mischief'.[5] In reply, a statement was issued by Hancock, Carter, Bunfield, Spencer and Varley describing the Mansfield circular as a 'cowardly and malicious attack on men who cannot retaliate without being accused of seeking to defeat national policy'.[6] A. J. Cook, in a blistering speech at Mansfield told the men that '. . . In this district you are like sheep without a shepherd, led to the slaughter' and he attacked both Spencer and Varley with fine impartiality.[7]

The result of the ballot was announced on Saturday, October 16th. On the first question, 14,331 were in favour of continuing the strike and 2,875 against. On the second question, 3,375 were in favour of negotiating a district agreement and 7,147 against. Half the members of the Association did not vote, and the vast majority of these were no doubt already back at work. It is also significant that whilst 14,331 members were prepared to vote in favour of the first proposition, with its emotional appeal to 'stand by the Federation Policy', only half that number voted against the positive suggestion that a district agreement—entailing a possibility of severance from the MFGB—should be negotiated.

On the strength of this result, Council resolved 'that all delegates who have returned to work, or have signed on for work, be now asked to leave the Council Room'. That was on October 16th. Three days later Council met again and on this occasion the officials of the MFGB were present. Council accepted the results of the ballot vote, and agreed 'to make every endeavour to carry into effect the General Policy of the Miners' Federation'. At the same meeting it was agreed that the resolutions passed at the previous Council meeting which suspended those delegates who had either gone back to, or had signed on for, work should be rescinded.

With the concurrence of the national officials, an attempt was now made to heal the breach with the Spencerite delegates, and indeed Cook in a speech at Eastwood invited Spencer himself to 'admit the error of his ways and come back'.[8]

But Spencer and the suspended delegates were powerless to get their men out again, although Cook's oratory had some temporary success. At Wollaton for example, only thirty men were at work on Monday, October 18th, compared with 600 the previous week. The effects of Cook's speeches soon wore off however.

At its next meeting, held on October 30th, Council resolved that the motion to rescind the suspension of delegates should not apply to J. B. Goff of Digby and H. Willett of Clipstone, whose sins were doubtless considered to be unforgivable. To make confusion worse confounded, Council then resolved once more 'that all delegates who have not carried out the Federation Policy be asked to leave the room'; whilst at the next meeting, held on November 6th, it was decided that this motion 'be not confirmed'.

The fact is that Council had lost control of the situation. It will be remembered that William Carter and George Spencer had expressed the view that to allow the substantial body of men who were back at work to remain unorganized was dangerous, and this certainly proved to be the case. Spencer claimed, in a House of Commons speech, that he did not start the breakaway but that this was already in existence before he arranged for the Digby men to return to work. This is undoubtedly true. There were indeed, all the makings of a breakaway from the very outset of the dispute: this is exemplified by the activities of Joseph Birkin the delegate who employed men on outcrop workings in defiance of the instructions of Council. Then in September and October, many lodges voted to return to work, and generally speaking, their delegates went back with them. Indeed, in some cases the delegates undoubtedly initiated the action which led to the decision to return to work. Well before October 6th negotiations were taking place between branch officials of the union and coalowners. There was, for example, a meeting at Edwinstowe between the Bolsover directors and a hundred branch committee members on 19th August.

By the beginning of October a state of chaos existed. Well over half the men were back at work, and a great many of those still out were anxious to return, In this sort of situation, an individual may find it extremely difficult to return to work, much as he might wish to do so. He may have a wife and family in dire distress, he may have exhausted his life savings and sold the greater part of his home; he may have pawned all his clothes and be faced with the possibility of eviction, but still he may find it difficult to go back. It is natural for a man to seek the approval of his fellows, and a man may be prepared to endure great physical suffering (not to mention his wife's nagging and the silent protest of his children) rather than commit an act which will lower him in the estimation of the community of which he is part. It may be however that if one person takes the lead, the rest may follow. The lead may be given by a union branch official consciously, but on the other hand it may be given by a non-conforming individual unconsciously. An example of the latter case comes to mind. A young miner who worked at Gedling Colliery and who supported a widowed mother and a younger brother decided to return to work. However, realizing his intention, a group of miners living in the neighbourhood forcibly restrained him. The next day several of these men themselves returned to work, and within a week they were all back.

Now, at the beginning of October a great many people were looking for a way out. The Federation policy was negative: they must not go back to work; further, the district leaders had no confidence in their policy and this was well known to the membership. It was obvious by this time that the original aims of the MFGB would not be fulfilled: defeat was inevitable. What then, was to be done? At some pits, men had trickled back, at others they had gone back in a body. Everywhere, strangers were taking the jobs of men still out. For those men who were working, there could be no bargaining with the employers. They took the terms which were offered and no questions asked.

And then along came George Spencer to arrange a return to work at Digby and New London. His action, being played out in an unstructured situation, at once assumed a wide importance. Immediately after the New London meeting he was approached by the checkweighman from Eastwood, and whilst he was eating his lunch at home, three more men came to see him. Before the day was out most of the pits covered by the old Nottingham and Erewash Valley Colliery Owners' Association had arranged to restart, and Spencer found himself thrust into the role of strike-breaker-in-chief.[9]

As we have seen, he was called to the National Conference to explain his actions, and was subsequently suspended by the NMA Council. Spencer felt that the condemnation he received from Herbert Smith and A. J. Cook was undeserved, and there is no doubt that this harsh treatment caused the breach to be wider and deeper than it need have been.

At the conclusion of the Conference at which Spencer was ordered out, the officers of the Miners' Federation came to Nottingham to try to get the men out again, and their efforts met with some temporary success. Spencer in a Parliamentary speech described what followed thus:

'After that, they had the council of action in Nottingham, and they dragged the men out again. Some of them have never returned to work since and upon my shoulders they are putting the blame for the last victimization. The blame should have been put upon the men on the council of action who foolishly came and dragged the men out again.'

However, whilst it is true that the Leen Valley men were brought out —largely by the oratory of A. J. Cook—they soon went back again. This time though, as George Spencer indicated, the owners refused to sign on the 'trouble makers', many of whom were to be permanently excluded from the industry.

At the Conference of the MFGB held on November 4th, William Carter reported that there were 40,000 men and boys' working and that many more would go to work if they could. Referring to the visit of the Federation Executive Committee, Carter said '. . . it is not propaganda we want, it is food we want, and it is starvation we are faced with. I know as good Trade Unionists as I am who have had to return to work, and had we been able to give money hundreds and thousands would have refrained from going back to work'. The following week, on November 10th, Carter reported that the numbers back at work had increased still further—to 44,000 out of a total labour force of 51,000.

The dissident delegates had by this time begun to take over the functions formerly exercised by the NMA Council under the leadership of George Spencer. In particular, they were negotiating a new district agreement to replace the temporary arrangements under which the men were then working.

The knowledge that this new organization, whose origins we shall discuss in greater detail in our next chapter, was now being licked into shape and that the owners were proposing to recognize it as the sole negotiating body for miners in the County increased still further the desire of William Carter and his colleagues for a speedy settlement of the dispute nationally. At the same time the bulk of the men who were now taking an active interest in the union's affairs were, as we have already remarked, left-wing militants who were still following an intransigent line. This gave a curious air of ambivalence to the Association's policy. Nationally, Nottingham's leaders opposed the South Wales policy of deepening the dispute, but the branches tended to support it.

At the Conference held on October 7th, Mr E. Morrell of South Wales moved the following proposals:

1. A return to the 'status quo' (i.e. the pre-stoppage pay and conditions).

2. The withdrawal of safety men.
3. An embargo on foreign coal (to be operated by the transport unions).
4. A cessation of outcropping.
5. The TUC to be asked to arrange for members of other unions to contribute to the miners' funds.
6. A propaganda campaign to be organized.
7. The whole of this work to be superintended and controlled by the Executive Committee of the MFGB.

This was seconded by Arthur Horner who asked whether: '. . . we are entitled to say that all our endeavours, based upon our desire to come reasonably to a settlement, have been interpreted as a weakening,' and who argued that the correct course to pursue was to do everything possible to starve the economy of coal, and then to wait for the other side to move.

Mr A. J. Cook, in an able and well-reasoned speech, opposed the South Wales motion which he characterized as unrealistic—as indeed, it was. Mr F. Swift of Somerset and Mr J. Baker of the Midland Federation both spoke of the impossibility of holding out much longer. For all that when the motion was put to the vote it was carried by 589 votes to 199—an overwhelming vote for a journey to cloud cuckoo land. Nottingham's twenty-five votes were recorded as opposing the South Wales Motion. However, at the same conference it was reported that the Government's offer of September 17th (described in our previous section) had been rejected by the districts, Derbyshire and Leicestershire being the only districts to vote in favour. On the face of it, we would have expected Nottingham's votes to be cast in favour of a return to work on the basis of district agreements (but with certain safeguards), whereas in the event they were cast against since the decisions taken at district level (i.e. as a result of branch votes) were being decided as we have seen, by the militant rump who were still on strike.

The South Wales resolution calling for an intensification of the struggle was referred to districts which endorsed it by 460,150 votes to 284,336. On this occasion Nottinghamshire and Leicestershire did not vote. The MFGB Executive Committee considered this result at its meeting held on October 15th when it was decided to consult with the National Federation of Colliery Enginemen on the calling out of safety men to ask the TUC General Council to assist in organizing an embargo on imported coal and to levy the affiliated membership for the benefit of the miners' relief fund, and to organize propaganda meetings throughout the coalfields. However, the impossibility of calling out the safety men, or of imposing an embargo on imported coal was manifest from the beginning. The safety men were not prepared to come out, and the transport unions were incapable of enforcing an embargo on imported coal even had they wanted. The levy proposal was a little more successful in that the TUC agreed to recommend affiliated unions to raise a voluntary levy. The Federation's propaganda also enjoyed a limited—and temporary—success in reducing the number of men at work in places like the Leen Valley. But there was no hope whatever of bringing the dispute to a successful issue by pursuing the South Wales policy. It was patent that British industry was receiving pretty well all the coal it needed; and that neither the Government nor the owners were prepared to make concessions.

In this situation the General Council of the TUC once more played an intermediary role. A TUC Council Sub-Committee met the Prime Minister on October 26th in order to explore the possibility of bringing the parties in dispute together. Three days later, on Friday, October 29th, the Sub-Committee met the Executive Committee of the MFGB and informed them 'that the position of the Government was, that it was prepared to set up a National Tribunal for co-ordinating district settlements, providing that the Federation agreed to recommend to its constituents that negotiations for such settlements should be opened up in the districts without delay'. The Executive Committee explained that they were still bound by the resolution of the last conference, but they agreed to call a further conference for November 4th when the whole matter would be discussed.

The conference duly met on November 4th and 5th. Some delegates believed, with Mr F. Swift of Somerset, that an immediate return to work should be arranged on the most favourable terms obtainable. Others, including the President Herbert Smith, thought that the Government's proposals for district settlements based on national principles should be thoroughly investigated. Others again thought with Arthur Horner that there should not be 'any change of policy which is three weeks old and which has brought definite and concrete and tangible results, and which I believe in a short time will bring us to a situation when the status quo will be possible and when these people will come forward with proposals much more palatable'. However, this view was not shared by Mr T. Richards, also of South Wales, who maintained that the three-weeks-old campaign to intensify the struggle had been a failure. Mr Richards warned the conference that unless a negotiated settlement were arranged quickly, a debacle would result, men would return to work without the protection of an agreement of any sort, and thousands of loyal members would be victimized. This was, of course, the argument used by Spencer a month earlier.

Nottingham's William Carter followed up on Richards's argument by pointing out that the NMA was almost on the point of collapse. Mr Spencer and the suspended delegates had agreed with the owners on the principle of a district settlement the result of which would be that the owners in future would ignore the Association and deal instead with the breakaway body. He appealed to the Executive to 'get the maximum offer with a view to saving the position, and submit it to the coalfield to see what the rank-and-file decide'. This appeal was echoed by Mr W. Straker of Northumberland, Mr P. Chambers of Scotland and Mr A. Bevan of South Wales. It was clear, from the views expressed by these—and subsequent speakers, that the overwhelming majority were in favour of exploring the Government's position with a view to negotiating an early settlement. Consequently, Conference adjourned for a few hours on the second day to enable the Executive Committee to meet the Prime Minister and other members of the Cabinet. At this meeting there was a certain amount of 'sparring' on both sides, but nothing definite emerged. It was agreed however, that negotiations should continue and that the National Conference should be recalled as soon as definite proposals had been formulated.

On the day following the adjournment of the conference, November 6th, the Secretary for Mines wrote to explain 'the general principles which the Government understands the owners in each district are prepared to follow in negotiating district settlements'. These general principles provided for a continuation of the system whereby wages were determined in each district by the results obtained in such district, the ratio for the division of net proceeds being within the range 87–13 to 85–15. It was also suggested that there should be a minimum percentage on basis equivalent to 20 per cent on standard[10] 'subject to district settlements of hours and working conditions'; and that subsistence wages should be agreed upon for the low paid day wagemen.

The Executive Committee of the MFGB, meeting on November 8th; felt that the Government's terms were defective in a number of respects. In particular, the Executive thought that the division of net proceeds between workmen and employers should be clearly laid down as 87–13; and that some 'national machinery for safeguarding national principles and co-ordinating district agreements should be established'. The Executive Committee also insisted that hours should not be the subject of district negotiations, but should remain at seven a day. At this stage the Executive Committee decided to reconvene the Special National Conference to hear a report on the situation.

The delegates to the National Conference reassembled on November 10th when A. J. Cook gave a full report. This was followed by a speech from the chair in which Herbert Smith urged that negotiations should continue despite the harshness of the terms on which it appeared a settlement would have to be reached. This speech drew a sharp protest from A. J. Cook.

Conference went on to receive reports from Districts from which it appeared that 237,547 men were at work. The apparent precision of this figure is, of course, deceptive: the press gave the numbers back at work as 313,000, which is probably a more reliable figure, so that something like a third of the normal labour force was working.

On its second day, November 11th, Conference had before it a Lancashire motion:

'That the Executive Committee be authorized to meet the Government to obtain the best terms possible for a settlement subject to any proposals being submitted to the members of the Federation by ballot vote before any settlement is arrived at.'

In moving this motion, Mr J. Tinker said that unless the Federation did something quickly, Lancashire could not be held firm. Mr Horner however, was still advocating a fight for the 'status quo', whilst Mr Cook made an impassioned speech in which he said 'We are now not even considering an honest settlement, we are considering surrender on the three points of principle for which for seven months the men have stood'. This speech was criticized by Mr F. Swift of Somerset who reminded Mr Cook 'that a month ago in this room he then declared that he could not possibly see any hope in continuing the struggle successfully'.

The majority view was summed up by Bob Smillie who said:

'We want to face the facts. . . . What is before us now is whether or not we shall give to the Executive power to tell the Government that we are willing to negotiate in districts, to negotiate as districts. . . . Now if we are facing facts, there is one fact, that we are getting weaker from day to day. It is stated that what we are discussing now is absolute surrender. It is not absolute surrender if you are going to take the form of giving power to negotiate terms locally.'

He went on to say, as George Spencer had said, that if an agreement were not reached the men would drift back to work without union protection and those who had been the most loyal would be victimized. Smillie was hopeful that if they could secure district settlements by district negotiations they would be able to rebuild the Federation, whereas failure to reach a settlement quickly might wreck the organization.

Eventually Conference decided by 432 votes to 352 to recommend the acceptance of the Government's terms as a basis of negotiation. (Nottingham, together with Derbyshire, Leicester, Forest of Dean, South Wales and Yorkshire, voted in favour of recommending the rejection of the terms.) However, when Conference's recommendation was voted upon in the districts, it was rejected by 460,806 to 313,200. This result is even more curious than it seems, since four of the districts which had opposed the terms at the conference (Nottinghamshire, Derbyshire, Forest of Dean and Yorkshire) now voted in favour of acceptance, whilst a fifth, Leicestershire, did not vote at all. The majority against now consisted, in addition to South Wales of the following districts: Cumberland, Durham, Lancashire, Northumberland, North Wales and Scotland, all of whom had supported acceptance of the Government's terms at the conference.

Nottingham's reversal was no doubt understandable. At the conference of November 10th to 13th, Frank Varley explained that Nottinghamshire would prefer district agreements to a national settlement involving reference to a tribunal, and it was presumably on these grounds that the county's vote was cast against the recommendation. The Nottinghamshire vote was not to be construed as opposing a speedy settlement; on the contrary a settlement had already been virtually reached between the owners and the Spencer Union, and Varley was afraid that the contrast between the terms that the Federation had been offered nationally, and the terms obtained by George Spencer locally would bring discredit on the Federation. However, when Branches voted on the matter—or rather when the militant group of the Branches voted—they indicated their continued support for a national settlement of some kind. This had, after all, been their raison d'être; and the Government's terms at least gave the appearance of a national agreement. Presumably, Yorkshire, Derbyshire and Leicestershire voted at the conference on much the same grounds as Nottinghamshire; whereas South Wales and Forest of Dean voted on Left Wing grounds: they were opposed both to district agreements and to the Government's terms.

At a further conference held on November 19th to consider the situation, Herbert Smith indicated that the Yorkshire delegates had taken back the majority decision of the last conference, and had asked their men to support it although they themselves were opposed to it. To use his own words:

'. . . so far as Yorkshire is concerned they cast a certain vote at this Conference and when this Conference decided with a majority against that opinion, Yorkshire went back as everybody else should have done and advised our people to accept these conditions.'

However, it was alleged that the Communist Party had organized a campaign opposing the conference recommendations, and A. J. Cook was supposed to have lent this campaign his support.

Mr Peter Lee said that in Durham a leaflet had been handed to the men attending Branch meetings which quoted a statement allegedly made by A. J. Cook in the *Sunday Worker* of November 14, 1926 which reads:

'My advice is to the men to reject the Government's terms. The TUC have acted as mediators for the coalowners urging us to accept first, lower wages, then longer hours, and now district settlements.'

The leaflet also contained a lengthy article by Arthur Horner and ended: 'Stand by Cook and reject the terms.'

Mr Cook repudiated the statement attributed to him, but Mr McGurk of Lancashire implied that Cook had opposed the Federation's policy in Lancashire during the previous week. This provoked the following reply from A. J. Cook: 'I object to the suggestion that I have been playing a double game. It is bad enough to be attacked outside without being attacked inside.'

The step finally taken by the conference of November 19th was dictated by the logic of events. It was resolved that districts should be recommended to enter into negotiations with the owners locally. The negotiations which followed however, were one-sided: the men were beaten, and their leaders had to accept whatever terms the owners chose to offer. Within a fortnight most districts were back at work. And so ended the most disastrous trades dispute in British history.

APPENDIX

TERMS FOR A RESUMPTION OF WORK

The following documents are taken from the files of The Digby Colliery Company.

The letter from the Digby Company dated August 20, 1926 indicates that the men were so demoralized by then that they were prepared to go back to work on almost any terms.

The Midland Counties Colliery Owners'
Association

Regent Buildings, London Road,
Derby.   19th August, 1926.

PRIVATE AND CONFIDENTIAL

Dear Sir,

Terms for resumption of work

In accordance with the decision arrived at by the Coal Owners of Derbyshire and Nottinghamshire at the Meeting held to-day, prints of the agreed terms are enclosed herewith, the intention being that one of these prints should be inserted in each of the Contract Books utilized by you, in order that men may be signed on under the agreed terms.

The copies enclosed herewith are designed to provide one for each of your Company's Pits, together with spare forms, which it is requested should be retained under lock and key, and no information regarding the terms made public in any way at the present time.

Yours faithfully,

William Saunders.
Secretary.

Terms for resumption of work, which Coal Owners of Derbyshire and Nottinghamshire are to be at liberty to offer to their workmen.

———

Hours of work per shift 7½.

Minimum percentage 36 on the 1911 basis.

Operating percentage for 7 months ending March 1927, 46·67.

Reduced percentage payable to piece-workers, 7·08.

That following the period of 7 months above referred to, the percentage for April 1927 be calculated upon the results of working in December 1926, and January and February 1927, and thereafter by a three monthly ascertainment.

The various provisions of the Agreement of 1921 to be operative in place of those of the 1924 Agreement, substituting for the provisions in the 1921 Agreement, as regards the division of profits, the proportions of 85 and 15 in place of 83 and 17. The Subsistence of 6d per shift to be continued within a maximum of 8/3.

Manipulators of Coal to work the Pit hours.

HB/EMC.

20th August, 1926.

Wm. Saunders, Esq.,
Regent Buildings,
London Road,
Derby.

Dear Sir,

Terms for resumption of work.

We beg to acknowledge receipt of your letter of yesterday enclosing six copies, which we note are to be taken care of and not left about or published. Our men sign on cards, one for each man, we are therefore putting one of the notices on the table, but our managers state that the majority of the men do not appear to take any notice of the details.

Yours faithfully,

THE DIGBY COLLIERY COMPANY LTD.
per

CHAPTER 10

THE 'SPENCER' UNION

The 'Spencer' union was by no means a mushroom growth. Long before 1926 there were people in the Nottinghamshire Miners' Association who were opposed to the increasingly militant attitude of the Miners' Federation. In the period so far covered in this volume (1914 to 1926) the locus of power inside the mining trade unions shifted from the district to the national level. National negotiating machinery replaced the district Conciliation Boards; and negotiations on fundamental issues took place not between district officials of the miners' and mineowners' organizations, but between the Executive Committee of the Miners' Federation on the one hand and the Central Committee of the Mining Association and the Government on the other. This development was inevitable during the war, but in the period which followed it the Miners' Federation attempted to convert temporary Government control into permanent national ownership of the industry, whilst the Mining Association sought to return to the district autonomy of the pre-war years. Politically, the Federation moved sharply to the left. Before 1914, most miners' MPs were Lib.-Labs. like Nottingham's J. G. Hancock, but as the old men died out or retired they were replaced by others who were free from the taint of liberalism. Further, with the winding up of the Federated District at the end of the war, the 'outside' districts (and, in particular, South Wales) played an increasingly important part in the Federation's affairs. This tendency was deplored by those who looked upon South Wales as the Syndicalist (and later, Communist) Trojan Horse.

We saw in Chapter 1, section 2, that the midland districts were still highly suspicious of the exporting districts in 1917 and 1918. Dealing with the South Wales policy of scrapping the district conciliation boards, George Spencer maintained that, whilst a national wages board might be a good thing, the owners would not be disposed to agree to it, so that unless the mines were nationalized . . . 'it would be a very unwise thing indeed to disband an institution (i.e. the Federated District) which has been so effective and so favourable to our own particular interests'.

In the period following the war it appeared for a time as though nationalization would be achieved fairly quickly, and in these circumstances even men like Spencer supported the idea of national settlements. However, it will be remembered that Spencer himself viewed the Coalition Government's apparent sympathy with the demand for Nationalization with scepticism: for example, he pointed out, on March 21, 1919 at a MFGB Conference, that Lord Sankey's interim report did not amount to a pledge that the industry would be nationalized. Once it became clear that the Government did not propose to nationalize the mines, Spencer concentrated without regret on industrial issues.

In the NMA there were those who believed that the union should play a full part in the political sphere; and there were those who preferred to refrain from political activity. The political section of the Association was predominantly left-wing whilst the 'anti-political' section was entirely right-wing. Indeed, the latter were not opposed to political action as such: they were opposed rather to politics of the wrong shade. Mr J. G. Hancock for example, had held office as a Lib.-Lab. MP and was opposed to the Association's political activities in the early 1920s because they were in support of the Labour instead of the Liberal interest. In particular, he was strongly opposed to the nationalization campaign. The same argument applies to Joseph Birkin, William Holland and the other adherents of the so-called British Workers' League and the Economic League which sought to convert the trade unions into 'non-political' organizations.

The struggle between left-wing and right-wing was polarized by A. J. Cook and his supporters on the Mansfield District Committee. This body attempted, during the period following the 1921 lockout, to win back to the union those men who had left it and they based their appeal on a left-wing programme. This brought them into conflict with the Associations' two MPs, Varley and Spencer, who possessed political complexions of too light a shade for the taste of their vocal constituents. These two men therefore, attracted the support of the 'non-politicals'.

The right-wing believed that A. J. Cook and his supporters had revolutionary aims; that is to say that they were not so much interested in ameliorating the miners' conditions as in changing the whole basis of society. Spencer it will be remembered, accused Cook of 'preaching revolution', and many of Cook's supporters were, of course, members of the Communist Party. It is a little difficult to know exactly what Cook believed since there was, as we saw in our last chapter, a great divergence at times between his impromptu open-air performances and his prepared Conference speeches. Cook was a highly emotional man who took upon his shoulders the weight of the centuries of injustice suffered by workers in general and miners in particular. At mass meetings of miners he found himself carried away by the fervour of his own oratory. His message was negative, but dealing as it did with the wrongs of the miners, it exercised a powerful appeal. Cook's prepared speeches were quite different. Their comparative moderation brought him into conflict with the Communists on more than one occasion. Nevertheless, Cook was widely regarded as the leader of the militant left-wing of the Federation.

Right at the very beginning of the stoppage, there was a powerful minority inside the NMA which opposed the Federation's policy. As they saw it, the Nottinghamshire miners had been made to stop work unnecessarily since their conditions were not immediately threatened. Further, many of those who supported the Federation at the outset turned against it towards the end when it appeared that the conflict was being prolonged not for economic reasons, since defeat was inevitable, but for political ends. This is true for example, of George Spencer himself since in the early stages, he supported the stoppage in obedience to the majority decision whilst by the time he arranged the return to work at Digby he could claim that he had a majority of the men behind him.

As we saw in our last chapter, the decision to call off the strike at Digby was taken on October 5th, when Spencer agreed to ask the management to allow all the men to go back to their old jobs. Within the next few days similar arrangements were made at Eastwood and at various pits in the area covered by the old Leen Valley owners' association. Spencer's expulsion from the MFGB Conference threw him into the arms of the owners; and the weeks which followed saw him organizing the breakaway body which had adopted him as its natural leader. Spencer threw himself into this work with a bitter intensity. His old political ambitions were now unlikely ever to be fulfilled; the career which he had mapped out for himself in the Labour Movement having been ruined, as he saw it, by the militants of the Left. By the beginning of November he appears to have made up his mind to establish a permanent union in opposition to the NMA. The owners, after a preliminary hesitation, agreed to give Spencer's Union sole negotiating rights and in return they expected labour relations to be maintained on a peaceful footing.

On Saturday, October 30th, it was widely reported in the press that a district agreement was to be made between representatives of the owners and of the men at work. On the same day, Council expelled nineteen delegates whose branches were at work. These delegates met in a separate room at the Miners' Offices and invited Mr Spencer to take part in a similar meeting to be held on Monday, November 1st.[1] This meeting was duly held, again at the Miners' Offices, under the chairmanship of Ben Smith. George Spencer was present during the whole of the meeting and F. B. Varley, J. G. Hancock and C. Bunfield attended for part of the time.[2] At this meeting it was decided, against the advice of Varley, Bunfield and Hancock,[3] to ask the officials to convene a meeting with the coalowners to be held on the following day. A meeting with the owners was arranged by Spencer but did not take place until the following week. Meantime, in an endeavour to save the situation, William Carter called a Special Council meeting for Tuesday, November 2nd to which all delegates including the Spencer Group were invited. The Spencer Group did not attend however, and in their absence Council adopted the following motion:

> 'Resolved that any Terms of District Settlement arranged by Mr G. A. Spencer and the suspended delegates with the local Coal Owners will not be recognized by the Nottinghamshire Miners' Association.'

Subsequently, Mr Varley's action in attending the meeting of the Spencer Group on November 1st came in for a good deal of criticism; indeed, it has recently been quoted to me as evidence of his implication in the Spencer breakaway. The account which Varley gave to the MFGB Conference on November 4th may therefore be interesting:

> 'They (i.e. the Spencer Group) fixed a meeting for Monday night, and I attended there and Mr Spencer, who was responsible for the official announcement, at the close of the meeting so worded his announcement as to make it appear as though I was identifying myself with this body of men, although the position was absolutely illogical; the Council was excluding men who were going to work and at the same time by resolution twice confirmed was accepting contributions from them. We went there. I told them exactly where they were going and in order to gain time until this Conference was held, asked them to agree to stand by the Federation in order to see what happened here, at any rate by some means. They said they were representing the majority opinion in Nottinghamshire, and would accept neither advice or anything else. The danger of the move arises from this fact: The meeting (with the owners) was to be held on Wednesday morning, and even whilst the Council assembled at 8 o'clock, we were informed that meeting was off. We were hoping there was sufficient humanitarianism in the coalowners to call that meeting off in the hope of negotiating something officially. We found that nothing had been done and the meeting takes place on Monday next.'

The owners were now ignoring correspondence from the Association.

Council, becoming increasingly worried about the way things were going, resolved on November 11th:

> '. . . that the General Secretary inform the Colliery Owners' Association that in the opinion of this Council the best interests of all concerned would be best served by deferring any Local Negotiations until the issues before the Government are made quite clear.'

This resolution was however, ignored in its turn.

Negotiations between the owners and the Spencer Group continued, and on November 20th an agreement was signed. This agreement, which will be found in an appendix to this chapter, gave the men an initial percentage on basis rates of 90 per cent which was to fall gradually to 60 per cent by June 1927. The percentage on basis rates prior to the stoppage was 46·67, and this had been increased to 56·67 in October 1926, so that the new agreement provided an immediate substantial increase in wages. For example, a face worker on day work with a basis rate of 8s 9d per shift would receive, in December 1926, 16s 7d against 12s 10d before the stoppage and 13s 9d in November 1926. This being so, Council saw no useful purpose in continuing the dispute and on November 22nd it requested all members to 'sign on' at the earliest opportunity.

On the night of the Council meeting, the Spencer Group held a meeting at the Victoria Hotel, Nottingham, where it was resolved that a new union, to be called 'The Nottinghamshire and District Miners' Industrial Union', should be formed. Mr Ben Smith, who had been the Vice-President of the NMA was elected President with Mr William Evans of Bolsover as the Vice-President. Mr Richard Gascoyne of Lowmoor was elected Secretary whilst Mr George Spencer was elected 'leader of the men at work represented'. It was decided that the owners should be asked to deduct union contributions of 1s a week for men and 6d for boys from wages and that a half of the monies so raised should be earmarked for a pensions fund.

On Monday, December 13th, the Executive Committee of the new union met representatives of the coalowners who confirmed that they would not recognize the NMA and who agreed to the setting up of a pensions scheme. On the question of deducting contributions from wages however, the owners explained that they were inhibited by the Truck Acts (although at a later date a form authorizing the deduction was devised and put into use).

The Pensions question was referred to again at a joint meeting held on Saturday, December 18th, when Mr H. E. Mitton stated, on behalf of the owners, that he had no doubt that the Nottinghamshire Coal Owners would agree to start a pensions fund by making an initial gift of £10,000. In return he made it clear that the owners would expect to have representation on the Pensions Committee. In a Parliamentary speech delivered in May 1927, George Spencer declared that this £10,000 was the only financial contribution made by the owners to the new organization; and that he had begged it from them for the sake of the retired miners who would otherwise have lost their pension rights consequent upon the depletion of the funds of the NMA. (This was not true since, according to the Spencer Union Minute Book a further gift of £2,500 was made in April 1927.) Mr Spencer undoubtedly set great store by the Pension Scheme which was evolved, and many old miners have reason to be grateful to him for it. However, we should not allow this consideration to blind us to the fact that the pensions scheme was, in its early days, used by the Industrial Union as a recruiting device.

At its meeting held on December 18th, the Industrial Union Council resolved:

> 'That all men who were drawing Old age pensions from the NMA or men at Bolsover and Creswell from the DMA before and during the Stoppage shall be eligible for Pensions from this Fund, also any Person who cannot now be restarted through Old Age be taken on.'

On March 5, 1927, it was resolved 'that in future when the NMA pay 2s 6d Pension, Secretaries to Branches of this Union shall only pay 2s 6d'; whilst on June 24th the Industrial Union Executive Committee recommended to Council that no more pensions should be paid to men whose sons were members of the Nottinghamshire Miners' Association. Thus was the old order of things reversed: the sins of the sons were now to be visited upon the fathers.

Despite the attractions of the pensions scheme the new union did not make nearly so much headway as it might have been expected to do, and the owners were therefore asked to render assistance in building up branch membership. Thus, on Saturday, December 18, 1926, the Industrial Union's Council instructed its secretary to write to the members of the Lodge and Cotes Park Collieries pointing out that the NMA were still being allowed to collect contributions at their collieries and that this was contrary to an undertaking given by the owners. This warning was ineffective and at its next meeting, Council decided to refer the matter to Mr W. Saunders (secretary of the Colliery Owners Association). At its meeting on February 5, 1927, the Executive Committee of the NMIU considered this matter and resolved 'that Mr G. A. Spencer and Secretary seek an interview with Mr H. Bishop, General Manager of the Manner's Colliery Company and that he be asked to assist the Representatives at the Lodge

Colliery to organize the Branch'. This meeting took place, and Mr Bishop undertook to stop the NMA people from collecting contributions at Lodge Colliery, but Council continued to protest that this undertaking was not being honoured. It appears that the Lodge Branch of the new union was not working at all satisfactorily since, at a later meeting, it was decided that the affairs of the Branch should be investigated and that its delegate, Mr Joseph France (who was also a member of the Executive Committee) should be suspended. The owners of Cotes Park, Pye Hill and New Selston (J. Oakes & Co.) continued to allow the NMA to collect subscriptions on the premises throughout the period of the 'split' and they were the only company to do so.

The union also complained to the owners about the unsympathetic attitude of colliery officials at the Kirkby Summit and Lowmoor pits where the NMA was still a force to be reckoned with, whilst the attitude of the management at Harworth to the branch delegate there was also the subject of a complaint in May 1927.

Meanwhile, the officials of the Nottinghamshire Miners' Association were attempting to bring life back into their organization. Their difficulties however, were many. Most of the active branch leaders were now out of work and were likely to remain so; the owners refused to negotiate with the Association so that if a man wished to have grievances rectified he had to use the services of the Industrial Union; and worst of all the Association were no longer allowed to collect contributions on the premises. Instead, in many cases contributions were collected in all weathers on some convenient street corner whilst in other cases this was considered to be unwise because of the possibility of victimization and collectors therefore went round from door to door. Matters were not helped by newspaper reports that J. R. MacDonald had written a letter of congratulations to Spencer on making a district settlement.

The Association was now back where it started. However, for a time its officials fought strongly for a rectification of the position. Its first duty was, of course, to protect the livelihoods of its members and for this reason members were advised to accept the Spencer terms for a resumption of work, but under protest. At the same time, the view was expressed that as soon as the members had had time to regain their strength, consideration should be given to tendering notices in order to force the owners to recognize the union once more. This is apparent in the resolution adopted by Council on November 27, 1926:

'Resolved as it now appears that the Miners of Nottinghamshire are not to have the Spencer Terms placed before them for accept-

O

ance, or otherwise, either by public meetings or by Ballot Vote, men are being pressed to sign the Terms in the Colliery office. If they refuse they may be dismissed or paid at the April rate of wages. Although the Spencer Agreement does not give in Wages what the Owners of Nottinghamshire could afford, although it is most distasteful to us, we could not advise men either to throw themselves out of work, or to deny themselves wages they are fully entitled to, we therefore instruct Miners as follows:

1. Vote against Mr Spencer negotiating for you.
2. Vote for securing the Official Agreement.
3. Refuse to sign for the Spencer Contribution being stopped at the Colliery office.
4. Accept any terms which give you more wages, as these can only bind you for the ordinary contract of service, viz. 14 days.

Finally, stick to your Union and refuse to be placed in bondage to the Employers.'

The Association issued 50,000 copies of a leaflet by Mr Varley in which the Wages Agreement negotiated by Mr Spencer was criticized, although this was a somewhat dubious manoeuvre since the Derbyshire Miners Association negotiated an agreement which was practically identical to the Spencer Agreement.

The MFGB Executive Committee took a serious view of the attitude of the Nottinghamshire coalowners. It decided on December 16, 1926, that the Federation should 'hold itself in readiness to do everything possible in the way of assisting the Nottinghamshire Miners' Association by way of meetings and propaganda work in the coalfield, as and when desired by Nottinghamshire'.

For the time being however, the NMA was left to fight its own battle; and whilst the owners were against it, it had the weight of tradition on its side. According to Frank Varley the actual membership of the NMA increased from 7,000 to 12,000 during the first fortnight of 1927 whilst the membership of the Industrial Union was estimated at 4,000 to 5,000. It is perfectly understandable that men who had been members of the NMA since boyhood should wish to rejoin it once they were settled in to work once more. On the other hand, even those who stayed out of the NMA would resent pressure being brought to bear on them to join the Industrial Union, and the bulk of them resisted this pressure for a long time. By June 2, 1927, the membership of the NMA had increased to about 13,500, despite the very many difficulties.

The greatest difficulty with which the Association was faced was, of course, lack of recognition. The Spencer Union had now formed, with the Owners, a district Wages Board and whilst the Enginemen's and Firemen's Union were allowed a seat on the Board, the NMA were rigidly excluded. At those pits where the checkweighman remained faithful to the NMA, the owners made it clear that he would only be allowed to continue in office so long as he refrained from conducting union business on the premises. For example, Mr Herbert Booth, the Annesley checkweighman, was approached by the colliery manager Mr Holt, in January 1927, and was told, as Mr Booth says,

'that the Industrial Union was the only Union they recognized and I should be allowed to continue as checkweigher so long as I kept strictly to the regulations of the Checkweigh Act. I was not to discuss any Union matters on the Colliery premises, nor to collect any Union dues, nor to post any notices of any character. If I did any of these things they would take steps to remove me.'

Quite a number of checkweighmen were put out of work: Arthur Thompson of New Hucknall, Alec Norris and Owen Ford of Welbeck and German Abbott and Tom Pembleton of Rufford among them. In these cases, the owners gave all the men notice to terminate contracts; then at the end of the notice period a few contractors belonging to the Industrial Union were re-engaged and they proceeded to elect new checkweighmen. The election of a new checkweighman automatically determined the employment of the old one who was thus forced out of the industry.

It is true, of course, that some of the checkweighmen belonging to the Industrial Union also found themselves in trouble with their employers (i.e. the workmen who were paid according to the mineral gotten); and two of them: W. G. Hancock and H. Peach of Bentinck, were actually thrown out of work.

It was by no means clear at this stage that the Industrial Union would win. Whilst it is true that they had the support of the owners, the NMA had a larger membership and enjoyed the passive support of the bulk of the non-members. However, the Industrial Union now campaigned strongly to carry the battle into the enemy's strongholds. Branches of the union were formed in various coalfields (the strongest outside Nottinghamshire and Derbyshire being in South Wales); and an attempt was made to win workmen in other industries over to the non-political trade union movement. An 'Industrial Peace Union' linking the non-political trade unionists with reactionary employers held conferences and issued statements designed to capture the allegiance of the rank-and-file. This development was viewed with concern by the General Council of the TUC who decided, as we shall see in our next chapter, that the best way to deal with this cancerous growth was to cut it out at its source.

APPENDIX 'A'

DOCUMENTS FROM THE FILES OF
THE DIGBY COLLIERY CO.

16th November, 1926.

To
Mr. H. Blundell.

NOTES OF A MEETING HELD AT VICTORIA HOTEL, NOTTINGHAM, BETWEEN NOTTS. COAL OWNERS AND WORKMEN'S REPRESENTATIVES known as the SPENCER GROUP

In opening the Meeting, THE CHAIRMAN (Mr. MITTON) said they were called together to discuss the question of a separate Agreement between the Owners and Men on the invitation of Mr. Spencer and his followers who claimed to represent the majority of the Men at present at work in the Notts. coalfield. The whole of the Owners in Notts. were represented and the majority of the pits were represented on the Men's side. All the Men's representatives present at the Meeting supported Mr Spencer's application for the Agreement. In reply to a question by the Chairman, MR. SPENCER said it was not possible to give a percentage of the Men at work he represented, but he felt he represented a considerable majority and pointed out that in response to a recent appeal by the old Notts. Miners Association support, only about 1,000 Men responded. He therefore claimed to represent more of the Men at work than any other side, THE CHAIRMAN said the Owners were as anxious as the Men to enter into an Agreement which would be fair and tend to a long period of peace in the district and suggested that whilst all the representatives present (both of Owners and Men) should be present at the deliberations of the Meeting, in order to facilitate progress, a Committee of 6 or 8 on each side should be appointed to restrict the discussion and ensure progress. Six on each side were duly elected, the Owners being represented by Messrs. Mitton, Phillips, Muschamp, Evans, Todd and Lees.

THE CHAIRMAN subsequently submitted various heads of Agreement most of which were accepted without much discussion, but others were left for further perusal at the Adjourned Meeting, viz.,

1. A District Wages Board to be constituted on equal representation. (Agreed to).
2. Wages payable to carry percentage on present basis rates, subject to a minimum percentage. (Agreed to).

3. Percentage to be paid and adjusted by results. (Agreed to).
4. Division of net proceeds in accordance with Government Memo. (Further discussion).
5. Minimum percentage to be agreed. (Agreed to).
6. Subsistence Wage to be paid to lowest paid Workmen. (Agreed to).
7. Hours of employment for those manipulating coal, $7\frac{1}{2}$ and $46\frac{1}{2}$. (Agreed to).
8. Provision to be made for recoupment by Owners on certain costs not provided for in the Ascertainment. (Further discussion).
9. District governing the Ascertainment to be agreed upon.
10. Schedule to be agreed as to items of expenditure to be included. (Further Discussion).
11. Period of first Ascertainment to be agreed upon. (Left Open).
12. Period of the Agreement. (Left Open).
13. Clause to confine present terms to March, 1927.
14. Reference to settlement in event of any question arising out of these terms.

MR. SPENCER elaborated on the present abnormal position of the Industry in Notts. and deprecated violent fluctuations in Wages. He advocated some system to keep Wages on a steady level, and suggested in the present case the Owners could be very generous and give the Men some present increase in Wages either by way of bonus or increased percentage. Men if generously met are prepared to enter into a long Agreement.

The Owners retired and after considerable discussion agreed to offer the following terms:

Terms on which Men returned to work be cancelled so far as Ascertainment of Wages percentage concerned, and the following substituted:

80% be paid for December, January and February next.
70% ,, ,, ,, March, April and May next.
60% ,, ,, ,, June, July and August next.

September percentage based on Ascertainment for May, June and July conditionally on acceptance in the Agreement of division of divisible surplus as to 85% and 15% with a minimum percentage of 36.

After consideration by Messrs. Spencer & Co., they suggested a further increase on percentage for first two periods and division of surplus in proportion of 86% and 14%. No agreement was reached and further discussion was left until (tomorrow) Wednesday and the Meeting was thereupon adjourned.

G. BURLINSON.

17th November, 1926.

To Mr. H. Blundell.

NOTES of continuation of a MEETING held at VICTORIA HOTEL, NOTTINGHAM, between NOTTS. COAL OWNERS AND WORKMEN'S REPRESENTATIVES, known as the SPENCER GROUP.

In opening the Meeting this morning, THE CHAIRMAN reviewed the proceedings of yesterday, presenting a copy to Mr. Spencer, who after perusing same pronounced it to be a correct report of the events which occurred. There was an addition to the numbers of the Men's representatives but two representatives of the Owners present yesterday, were absent to-day (Mr. Oakes and Mr. Todd Junr.) THE CHAIRMAN then reviewed the position as left yesterday where the chief difficulty centred around the minimum precentage of 36% and the division of the Ascertainment Surplus as to 85% and 15%. He invited a review of the position of the Notts. Collieries during the nine months preceding the Stoppage when the Ascertained percentage was below the minimum in every month but on account of the pits being 'spoon fed' by the Government the minimum percentage of 46·67% was paid. He impressed on the Meeting the sad plight of the pits without this assistance and emphasized the need for the minimum being reduced to 36%, laying particular stress upon the unknown effects that would arise if other districts settled on an eight-hour basis and the increased outputs that were likely to result. MR. SPENCER replied at length and stressed upon the result that would accrue if Ascertainments arrived at on basis of December results. He advocated a levelling of wages as against great fluctuations and pleaded for some consideration to be shewn immediately to the Men in recognition of the unexpected results which had already accrued to the Owners through the Notts. pits being the only ones at work. He gave figures from Government returns shewing that division of Surplus as to 87 and 13, gave a very favourable return to the Owners—for a period of months this shewed a profit of 1/2d. per ton. He pleaded hard for division on basis of 86 to 14, and evidenced figures quoted as shewing reasonableness of his claim.

The Owners then retired and considerable discussion ensued continuing late into the afternoon on the question of the Minimum Percentage and the ratio of division of ascertainable Surplus. It was ultimately decided to offer the Men the following terms, viz:

That the December Percentage be 90% instead of 46·67%.
„   „   January and February, 1927 be .   . 80%.

That the March, April and May, 1927 be .   . 70%.
,,   ,, June, July and August, 1927 be .   . 60%.
,, September be on Ascertainment of May, June and July.
,, Surplus be divided as 85% to Wages and 15% to Profits.
,, The Minimum Percentage be 38% on basis rates with a
     recoupment clause to be operative over a balancing period
     of 12 months.

After further discussion with the Men, these terms were ultimately
accepted and discussion then centred around the Subsistence Wage.
The Owners suggested doing away with the Percentage and substituting a maximum of 8/3d. with a minimum of 7/6d. for able-bodied
Men. No Agreement was reached and further discussion was left
over until to-morrow (Thursday).

G. BURLINSON.

18th November 1926.

To Mr. H. Blundell.

NOTES of continuation of MEETINGS held at VICTORIA
HOTEL, NOTTINGHAM, between NOTTS. COAL
OWNERS AND WORKMEN'S REPRESENTATIVES,
known as the SPENCER GROUP.

Continuing the proceedings of the past two days, the Meeting this
morning commenced by the Secretary reading a resume of the events
of yesterday. This was agreed to generally by Mr. Spencer who
however, queried the interpretation of the divisible surplus as to
85 and 15, and this question occupied the larger part of the deliberations throughout the day and the Meeting ultimately closed without
any agreement being reached on this question. Before the first
adjournment by the Owners, THE CHAIRMAN informed the
Meeting that he had received a request from Mr. Hall of the Derbyshire Miner's Association, asking the Owner's Association to meet
the joint representatives of the Notts' and Derbys. Miners' Union
for the purpose of entering into an Agreement. Several Miners'
delegates who had attended all the Meetings at Basford said this
request was quite unauthorized so far as the Notts. Men were concerned as the question had never been put before the delegates. THE
CHAIRMAN said that the Owners having decided to meet Mr.
Spencer and his followers would not meet anyone else without Mr.
Spencer and his friends and that the proceedings already instituted
would be proceeded with to a finish. MR. SPENCER thanked the
Owners for this decision and read a telegram from Mr. McDonald
in which he said he hoped Spencer would flourish in spite of every-

thing. The Owner's representatives then adjourned to consider the question of the Subsistence Wage. After lengthy deliberation, it was decided that Clause 6 governing this subject be included in the Agreement as follows:

'*Clause* 6   To any adult able-bodied day wage workman whose rate per shift, after adding district percentage, is less than 8/9d. a subsistence wage shall be paid sufficient to bring him up to 8/9d., provided that the maximum addition in any instance shall not exceed 6d. per shift. This wage applies only to ordinary working time.

'Provided also that no adult able-bodied day wage workman shall be paid a gross rate, including the subsistence wage, of less than 7/11d. per shift on his ordinary working time, not including overtime.'

This was accepted by the Men's delegates.

The Owners' representatives then went into Committee on Clause 11 which was ultimately settled as follows:

'The Basis wages payable by the Owners for the months of December, 1926 and until the end of August, 1927 shall carry percentages as follows: viz:

> December 1926   90% instead of 46·67%
> January and February 1927   80%
> March, April and May 1927   70%
> June, July and August 1927   60%
> September to be on ascertainment of May,
>   June and July 1927

The wages payable by the Owners during any month afterwards shall be based upon the ascertained results of the first three of the four months immediately preceding the month of ascertainment unless and until the periods of ascertainment thereafter are varied by the Board.'

Clause 12 was next considered and it was decided to recommend the period of this Agreement to extend over five years from 1st December, 1926 and be determined afterwards by six months' notice from either side. The period was agreed to by the Men's representatives but the question of the period of notice was left for further discussion.

Clause 8 re division of proceeds was next reconsidered, but without any agreement being reached and the Meeting was ultimately adjourned until Saturday morning next when this matter, also the third Schedule of the 1924 Agreement would be further considered.

G. BURLINSON.

APPENDIX 'B'

AGREEMENT
made between the
NOTTINGHAMSHIRE AND DISTRICT COLLIERY OWNERS
and
REPRESENTATIVES OF THE WORKMEN
working at Nottinghamshire and District Collieries.

1. A District Wages Board (hereinafter called 'the Board') shall be constituted consisting in equal numbers of persons chosen by the coal owners in the district and persons who are employed or have been employed at Collieries in the District chosen by the workmen employed at the Collieries in the District. The Board shall draw up its own rules of procedure which shall include a provision for the appointment of an independent Chairman.

2. The wages payable in the District during any period shall be expressed in the form of a percentage of the basis rates then prevailing and shall be periodically adjusted in accordance with the results of the industry as ascertained in the District.

3. The amount of the percentage to be paid in the District during any period shall be determined by the results of the industry in the District during a previous period (hereinafter called the period of ascertainment) as ascertained by returns to be made by the owners, checked by joint test audit of the owners' books carried out by independent firms of Accountants appointed by each side.

4. Subject to the provisions of Clause 8 hereof, in order to determine the percentage payable in the district in accordance with Clause 3, 85 (eighty-five) per cent. of the difference between the proceeds and the costs of production other than wages in the District during the period of ascertainment shall be taken. From the amount so determined shall be deducted the amount paid during the like period as allowances under Clause 6, and the balance so remaining, shall be expressed as a percentage of the wages paid at basis rates during the period of ascertainment.

5. Wages in the District shall not be paid at lower rates than basis rates plus 38 (thirty-eight) per cent. thereof.

6. To any adult able-bodied day wage workman whose rate per shift, after adding district percentage, is less than 8/9 (eight shillings and ninepence), a subsistence wage shall be paid sufficient to bring him up to 8/9 (eight shillings and ninepence) provided that the maximum addition in any instance shall not exceed 6d. (sixpence) per shift. This wage applies only to ordinary working time.

Provided also that no adult able-bodied day wage workman shall be paid a gross rate, including the subsistence wage, of less than 7/11 (seven shillings and elevenpence) per shift on his ordinary working time, not including overtime.

7. The hours of work per shift for manipulators of coal above and below ground shall be half an hour longer than those in operation in the District above and below ground respectively on April 30th, 1926. The percentage payable to piece-workers shall be 7 (seven) per cent. in place of 14·17 per cent.

8. If the amount of the ascertained proceeds in respect of any period of ascertainment is less than the sum of the amounts of (1) costs other than wages (2) the cost of wages at the minimum rates as provided for in Clause 5, and of allowances paid during such period under Clause 6, and (3) an amount equal to 15/85ths (fifteen-eighty-fifths) of the cost of such wages and allowances, the deficiency shall be carried forward and dealt with in subsequent periods of ascertainment according to the following method so as to secure effective recoupment thereof:

In any ascertainment in which the amount of the proceeds is greater than the amount required to meet (1) costs other than wages (2) the cost of wages at the minimum provided for in Clause 5, and of allowances paid during the period of ascertainment under Clause 6 hereof and (3) an amount equal to 15/85ths (fifteen-eighty-fifths) of the cost of such wages and allowances, the balance shall be applied so far as may be necessary to make good any deficiency brought forward from previous ascertainments. Of that part of the balance which remains after meeting the deficiency, 85 (eighty-five) per cent. shall be applied to wages. If there be no balance available for meeting a deficiency or if the deficiency brought forward exceeds the balance, the deficiency or such portion thereof as remains shall be again carried forward to be made good in a subsequent period or periods according to the above method.

Provided that no deficiency shall be carried forward for recoupment purposes beyond the 30th April in any year.

The first ascertainment period shall be May, June and July, 1927.

9. The District shall comprise Nottinghamshire and District, and shall only be varied by the decision of the Board.

10. In ascertaining the proceeds, wages and costs of production

other than wages, the Accountants shall follow the principles set out in the Schedule hereto, and any amendment or addition to such principles which may hereafter be adopted by the Board.

11. The basis rates payable by the owners for the months of December, 1926, and until the end of August, 1927, shall carry percentages as follows:

> 90 (ninety) per cent. on basis rates
> for December, 1926.
> 80 (eighty) per cent. on basis rates
> for January and February, 1927.
> 70 (seventy) per cent. on basis rates
> for March, April and May, 1927.
> 60 (sixty) per cent. on basis rates
> for June, July and August, 1927.

The wages payable by the owners for the month of September, 1927, shall be based upon the ascertained results of the months of May, June and July, 1927, those for the month of October upon the ascertained results of the months of June, July and August, and so on, unless and until the periods of ascertainment thereafter are varied by the Board.

12. The period of the duration of this agreement shall be to the 31st December, 1931, and thereafter until terminated by 6 (six) months notice on either side given on the first day of any month following.

Dated this 20th day of November, 1926.

(Signed) For and on behalf of THE COAL OWNERS	(Signed) For and on behalf of THE WORKMEN
H. Eustace Mitton. (Butterley Co. Ltd.)	Geo. A. Spencer.
C. W. Phillips. (Barber, Walker & Co. Ltd.)	Wm. Evans (Bolsover).
P. Muschamp. (New Hucknall Colly. Co. Ltd.)	B. Smith (New Hucknall).
	Horace W. Cooper (Creswell).
R. H. F. Hepplewhite. (Bestwood Coal & Iron Co. Ltd.)	Jas. France (Lodge Colliery).
	Haydn Green (Brinsley).
T. A. Lawton. (Babbington Coal Co.)	Richard Gascoyne (Lowmoor).
Hubert O. Bishop. (Manner's Colliery Co. Ltd.)	William Buckley (Wollaton).
	Joseph Cobley (Clifton).
	H. Willett (Clipstone Colly).

(Signed)
For and on behalf of
THE COAL OWNERS

S. Evans.
(Bolsover Colliery Co. Ltd.)

W. Dawson.
(Wollaton Collieries Co. Ltd.)

H. N. Berry.
(Clifton Colliery Co. Ltd.)

Chas. W. Gray.
(Linby Colliery Co. Ltd.)

W. H. Mein.
(Sth. Normanton Colly. Co. Ltd.)

Geo. Burlinson.
(Digby Colliery Co. Ltd.)

Chas H. Heathcote.
(Sherwood Colliery Co. Ltd.)

C. R. Ellis.
(Blackwell Colliery Co. Ltd.)

P. F. Day.
(Pinxton Collieries Ltd.)

N. D. Todd.
(Stanton Iron Works Co. Ltd.)

W. G. Ball.
(James Oakes & Co. (Riddings
Collieries) Ltd.)

Alfred Hewlett.
(Cossall Colliery Co. Ltd.)

Thomas G. Lees.
(Newstead Colliery Co. Ltd.)

(Signed)
For and on behalf of
THE WORKMEN

T. Lake (Clipstone Colliery).

Ernest Lenthall (Bolsover
Colliery).

Nathaniel Buxton (Gedling).

Tom S. Willoughby (Broxtowe).

Joseph Wilson (High Park).

James Benjamin Goff (Digby).

Joseph Shooter (Rufford).

Aaron Jenkins (Ollerton).

John Hackett (Trowell Moor).

Edwin Lee (Cotes Park).

Jos. Birkin (Moor Green).

Arthur Spencer (South
Normanton).

James G. Sears (Bestwood).

Fred E. Albon (Hucknall No. 2.)

Vin. Goddard (New London).

William Gervase Hancock
(Bentinck).

James Shaw (Linby).

Witness to all the foregoing signatures:

William Saunders,
Regent Buildings,
London Road, Derby.

CHAPTER 11

THE TUC INTERVENES

A special conference of the MFGB met on June 2, 1927 at the Kingsway Hall, London. At this Conference Mr Hicken of Derbyshire complained that at the pits belonging to the Bolsover Colliery Company, the Spencer union was now recognized by the owners to the exclusion of the Derbyshire Miners' Association. The Industrial Union was also operating in South Wales (where it was known, according to Mr E. Morrel, as the 'Government's Club'); and in South Yorkshire, where the strongest branch was at Barber and Walker's Bentley Colliery.

In South Wales, Spencer had a powerful ally in the Powell Duffryn company who found the Industrial Union's solicitor an easier man to deal with than the officials of the South Wales Miners' Federation. However, Mr Spencer's most powerful ally was Mr Havelock Wilson of the National Union of Seamen, who took upon himself the generalship of the non-political trade union movement.

Joseph Havelock Wilson was born in Sunderland in 1859.[1] He went to sea at an early age, and became interested in Trade Unionism whilst sailing in Australian waters. Wilson joined the Australian Seamen's Union and upon returning to this country he joined a local Seamen's Union centred on Sunderland. In 1886 Wilson's wife and her relatives opened a restaurant and Wilson himself left the sea in order to help run the business and to concentrate on Trade Union work.

Wilson decided to start his own Union which would endeavour to recruit members nationally. At the inaugural meeting of the new Union, only two people, including Wilson, attended. The Union was given the grand title of: 'The National Amalgamated Sailors' and Firemen's Union of Great Britain and Ireland.' This body, faced with financial difficulties went into liquidation in 1894; and Wilson then formed the 'National Sailors and Firemen's Union'. For a very long time the employers refused to recognize this union but during the first World War a joint body was formed from which developed in 1920 the National Maritime Board, whose purpose it was to settle all disputes peacefully. Eventually, the rival Seamen's Unions disappeared, after some very underhand work on the part of Wilson and his henchmen,[2] and the employers helped Wilson's Union (now called the National Union of Seamen) to organize all seafarers.

The National Union of Seamen was Wilson's creation; and he would brook no opposition to his plans. A seamen's union is, of course, difficult to organize unless (as in the case of the NUS) the employers agree to maintain a 'closed shop'. However, given the co-operation of the employers, such a Union can easily be dominated from its head office since branch life is necessarily at a low ebb when members are scattered about the globe. Even when sailors are ashore they have affairs to attend to which will doubtless appear to most of them to be of far more pressing importance than attending branch meetings. The National Union of Seamen in Havelock Wilson's day was therefore a client union. Havelock Wilson, as Mr Shinwell says in a letter to the author, 'made a bargain with the shipowners so that nobody outside his union should be employed, and opposed the Labour Party in every way'. Like J. G. Hancock and George Spencer, Wilson had been a Member of Parliament but after losing his own seat he became an advocate of 'non-political' trade unionism.

When the General Strike was embarked upon, Wilson's Union was alone in opposing the decision to strike. Then when the Spencer union was formed, Wilson took it under his wing and tried to widen the movement so as to make it a national organization of people in various trades. In practice, 'The Non-Political Trade Union Movement' achieved scant success outside the Nottinghamshire coalfield; but this was not Wilson's fault. He threw himself into the fight with his customary energy, and his NUS organizers went round the country trying to arouse enthusiasm for the anti-socialist crusade.

On July 28, 1927, a deputation from the National Union of Seamen, consisting of Messrs Cotter, Davies and Bond[3] attended a National Conference of the MFGB. They explained that, at a recent meeting of the Executive Committee of their Union it had been decided to make an interest-free loan of at least £10,000 to the Spencer Union. However, Mr Davies who was the General Secretary of the NUS, had refused to sign the cheque, and he and his two companions had decided to test the legality of the proposed loan. Havelock Wilson, as the President of the National Union of Seamen, then called a Special General Meeting of the Union which was held on August 1st and at which the decision to lend money to the Spencer Union was confirmed. Subsequently the officials who had opposed this move were expelled from the Union.

The MFGB referred the matter to the TUC General Council. They had a number of complaints in addition to the one about the proposed loan of not less than £10,000: They complained, for example, that the journal of the Miners' Non-Political Union was issued from the headquarters of the NUS, that meetings in support of the breakaway were being held under NUS auspices, and that the NUS had

placed three or four organizers, with cars, at the disposal of the
Spencer Union in South Wales. At a dinner on July 8th at the headquarters of the NUS attended by representatives of the Seamen's and
Spencer Unions, together with Mr Frank Hodges, Havelock Wilson
was reported to have said:

> 'They had decided to establish an industrial union for the miners,
> in which politics would be debarred. He heard with some disgust
> from a few that if the Miners' Industrial Union was helped it
> would mean supporting a "breakaway" union. . . . It was going
> to be a big fight, and he had been warned of the terrible conse-
> quences that he would suffer at the hands of the Trades Union
> Congress. He was not dismayed in the fight for freedom and
> sound Trade Union principles. He intended to back the miners
> right up to the hilt, and in that he would go "the whole hog".'[4]

For a meeting held at Newcastle, at which Hodges, Wilson and Spencer were billed to speak, colliery officials issued employees with free
railway and bus tickets to induce them to attend. Other workmen
were taken to the meeting by car. On October 26, 1927, the General
Council of the TUC, acting on the report of its Disputes Committee,
called upon the National Union of Seamen to:

> '1. Refrain from implementing the promised loan of £10,000
> decided upon at the Special General Meeting of the union held on
> the 4th August, 1927, and to cease forthwith from supporting or
> encouraging financially, morally, or in any other way, the Miners'
> Industrial (Non-Political) Union.
> '2. Give within 14 days a written undertaking to refrain from
> supporting or encouraging financially, morally or in any other way,
> the Miners' Industrial (Non-Political) Union or any union or
> unions connected therewith either directly or indirectly.'

The National Union of Seamen refused to give the undertaking
asked for, and the General Council therefore took steps to have the
Union excluded from the TUC.

In the early part of 1928, the General Council joined with the
MFGB in arranging an intensive propaganda campaign in the Nottinghamshire coalfield. Week-end meetings, with well-known speakers
including the veterans Will Thorne and Ben Tillett, were held at
various points in the County and in the Bolsover-Creswell district in
Derbyshire. The General Council also agreed to render assistance to
the Nottinghamshire Miners' Association in a dispute at Welbeck
Colliery. Here the owners had given all the men notice thus bringing
the employment of the two checkweighmen, Owen Ford and Alec
Norris, to an end. During the notice period reduced price-lists had

been agreed with the Spencer Union and only those men who would agree to accept the reduced prices and also to have contributions to the Industrial Union stopped out of their pay were to be re-employed. The locked-out men were supported by grants made by the MFGB.

In March 1928, a joint meeting of the MFGB Executive Committee and a sub-committee of the TUC General Council, meeting in Nottingham, agreed that the TUC should bring pressure on the Nottinghamshire coalowners to stop the discrimination in favour of the Spencer Union. The Coalowners' Association agreed to meet the TUC sub-committee and at this meeting, held on March 21, 1928, they denied that pressure was brought to bear on men to join the Industrial Union, but they maintained that Spencer's Union had a larger membership than the NMA and was more representative of the men and they therefore proposed to continue to allow it sole negotiating rights. The Owners Association agreed to issue a public statement outlining their attitude and this was given wide publicity by the TUC. The statement reads as follows:

'The Nottinghamshire Coal Owners' Association intend loyally to carry out the agreement made with representatives of their workmen, and cannot recognize any other organized body of workmen than that with which they are now dealing. The Nottinghamshire Coal Owners' Association however, make no inquiry or discrimination in regard to a man's trade union.'

Despite this statement, victimization of NMA members continued at many pits. The TUC General Council considered that the owners' assertion as to Spencer's popularity should be tested by a secret ballot and, by agreement with the MFGB and the NMA, the ballot was held on April 30th and the succeeding days. The ballot was under the general supervision of a reputable firm of solicitors, Eking, Manning, Morris and Foster, and every endeavour was made to ensure that the vote was properly taken and counted. Attempts were made by officials of the National Union of Seamen to bring the proceedings into disrepute, but without success. Further, some branch officials of the Industrial Union attempted to intimidate voters. Despite all the difficulties, the ballot was successfully organized and when the votes were counted on May 4, 1928, it was found that 32,277 men had voted in support of the NMA against 2,533 votes cast in support of the Industrial Union.

This did not however make any difference to the attitude of the Nottinghamshire Coal Owners and the MFGB therefore decided to organize a further series of week-end meetings in the coalfield commencing on June 8th. A personal canvass of miners, with a view to building up the membership of the NMA, was conducted and both

Herbert Smith and A. J. Cook took part in this despite their many other duties.

Meantime, the victimized miners of Welbeck—800 in number—together with the two checkweighmen were unable to obtain work, their places having been filled by others. They received dispute pay provided by the TUC and the MFGB for eight weeks, but on May 9th, A. J. Cook had the unpleasant duty of informing them that the payments would no longer be made after that day. Thereafter they had to make do with the 'dole'.

The claim of the owners that there was no discrimination against members of the Nottinghamshire Miners' Association convinced nobody. Even so liberal an employer as the Digby Colliery Company, belonging in the main to the Bayley family, was said by Herbert Smith to allow Industrial Union men to have the best districts. Further, the man who was in charge of the men's Pension Fund was told that he must not go on the premises. The Digby situation was doubly annoying to the Miners' Federation first because Sir Dennis Bayley had always been considered one of the union's best friends and second because the manager of the pit was the son of J. G. Hancock. In fairness to the Digby Company however, it must be admitted that their policy was still liberal by comparison with that of the majority of the owners, particularly those whose pits were in the newly-developed part of the coalfield around and to the North of Mansfield. In some mining villages, indeed, to be known as a member of the NMA was to invite dismissal.

To help recruit new members and to encourage old members to remain loyal, the TUC agreed to pay eight collectors three day's pay a week to conduct a regular house-to-house canvass. This system was continued by the NMA and remained in force until 1937.

The TUC Sub-Committee responsible for securing an improvement in the position in the Nottinghamshire Coalfield continued to meet for some little time, but its efforts were unavailing. There appears to have been a general feeling among the leaders of the TUC that if the 32,000 Nottinghamshire Miners who wanted to join the NMA would only have the courage of their convictions the difficulty would be solved overnight; and because of this feeling the TUC lost interest in the issue. It must have seemed incredible to people in other industries that so many grown men should allow themselves to be terrorized by so few. Victimization there was, of course, in plenty. For example, the MFGB Executive Committee had before it in July 1929 a report which stated that two men who had been trying to form a branch of the union at Bilsthorpe Colliery had been dismissed, 'ostensibly because of trade depression, but actually because of their activities on behalf of the Nottinghamshire Miners' Association';

P

and that at Annesley Colliery, 200 men had been given notice and had been notified that the notices would operate unless they agreed to join the Spencer Union. However, had the general body of the men been sufficiently determined, victimization of this kind could not have taken place.

In 1929 therefore, the TUC talked about the problem and did nothing more. In September 1930, to show its solidarity with the Nottinghamshire miners, the TUC held its Annual Congress in Nottingham. But the coalowners remained unmoved; and George Alfred Spencer took no notice, and the miners of Nottinghamshire cursed both the coalowners and Spencer but continued working just the same.

Whilst it may be true, as the left alleged, that the TUC failed to press this matter with sufficient vigour; it is equally true that the indifference of the average Nottinghamshire miner gave it very little encouragement.

CHAPTER 12

MONDISM AND THE MINORITY MOVEMENT

During the General Strike the Government maintained that the Strike was illegal. When the Strike was over however, they decided to make assurance doubly sure by imposing new legal restrictions on the Trade Union Movement.

The Trade Disputes and Trade Unions Act of 1927 made sympathetic strikes involving people in more than one industry illegal. Similarly, strikes designed to coerce the Government either directly or by causing hardship to the public were also declared illegal. The old offence of 'intimidation', which had been used against the unions in Victorian times was revived and the restrictions on picketing were tightened up. Amongst other provisions of the Act was one which altered the law governing the payment of political levies. Under the 1913 Act, any member of a union with a political fund was under an obligation to pay political levies unless he signed a 'contracting out' form. Under the new Act the procedure was reversed: a man now had to sign a 'contracting-in' form if he wished to pay political levies.

The unions feared that the new law would cripple their activities, but in practice they suffered very little from it. Although the instrument of repression was at hand the Government chose not to use it. There was, indeed, very little need to use it. The General Strike—and the miners' lockout which followed it—was a salutary lesson to both sides. The employing class, and the Government which was primarily representative of their interests, had to recognize the potential power of the trade union movement. And the trade union leaders for their part recognized that in any future large-scale dispute they might be carried forward by the logic of events into an outright revolution. There was a widespread feeling on both sides that open conflicts were wasteful and unnecessary. And so we find a marked diminution in industrial disputes in the period which followed. Francis Williams points out that between 1919 and 1926 an average of over forty million working days were lost through strikes and lockouts each year; whereas in the seven years following the General Strike on an average only four million working days a year were so lost. This could possibly be accounted for in part by the weak bargaining position of the unions during a period of heavy unemployment except that the anti-militant tendency was intensified during the upswing of 1935-7. Further, we cannot account for the

Anti-Socialist & Anti-Communist Union   New Series No. 116.

The
Trade Unions Bill

Vindicated by a
Labour M.P.

*(From the speech delivered by Mr. G. A. SPENCER,
M.P. for Broxtowe, in the House of Commons, May
2nd and 3rd, 1927).*

ONE PENNY.

Special quotations for quantities.

Issued by

THE ANTI-SOCIALIST & ANTI-COMMUNIST UNION,
(Founded 1908)

58 & 60, Victoria Street, London, S.W.1.
TELEPHONE: VICTORIA 6710.

" We Educate."

5. Front page of reprint of George Spencer's defence of the
1927 Trade Unions Bill

tendency by supposing that the trade unions were licking their wounds preparatory to a renewal of conflict. Fortunately, there is no need for speculation on this subject since the facts are clear. The trade union movement, under the leadership of Walter Citrine and Ernest Bevin, deliberately turned their backs on their militant past and concentrated instead on winning what concessions they could by peaceful means. The employers for their part, except in the mining industry, accorded the unions a much wider measure of recognition than ever before. Indeed, 'recognition' disputes accounted for less than 3 per cent of disputes in the seven years following the General Strike compared with, for example, 30 per cent in 1923.[1]

The Non-Political Trade Union Movement.

(Founded by the late J. HAVELOCK WILSON, C.H., C.B.E.)

Secretary and Financial Organiser: F. H. PRUEN, M.A.	President: GEORGE A. SPENCER, Nottingham	Telephone: LONDON WALL 1328 & 1329.
City Representative: MRS. E. A. CHAPMAN.		3, LONDON WALL BUILDINGS,
All Cheques to be made payable to: THE NON-POLITICAL TRADE UNION MOVEMENT.		LONDON WALL,
		LONDON, E.C.2.

6. Letter heading of the Non-Political Trade Union Movement

In the mining industry, of course, a very serious attempt was made to break up the Miners' Federation of Great Britain. It is true that the Spencer Union made far less headway outside Nottinghamshire than once seemed likely, but this was not for want of trying on the coalowners' part. Branches of the Spencer Union were fostered in all the coalfields—there were said to be 273 in 8 counties in 1928—but (with the exception of one district of South Wales) they did not take root. This is understandable. Once the miners were back at work their unions were able to renew their strength. By the time the Industrial Union entered its missionary phase the heathen had become confirmed in his error. Had the Great Lockout lasted a few months longer, the result might have been very different.

If the attempt to form a national Spencer Union in the mining industry was a failure, the attempt to establish a non-political trade union centre in opposition to the TUC was doubly so. The Non-Political Trade Union Movement with its headquarters in the City could, of course, boast the support of the National Union of Seamen so long as Havelock Wilson was alive, but with his death the Movement did little more than provide George Spencer with an impressive London address.

The employers could then, as a class, have used the instrument of repression provided by Parliament in 1927; or they could have relied upon the attempt to foster breakaways on the Spencer pattern, In fact, they chose to do neither. Instead, they formed a compact with the official trade union movement which has come to be called 'Mondism' after Sir Alfred Mond (who became Lord Melchett), head of Imperial Chemical Industries.

Mondism was a policy of co-operation between the two sides of industry in the interests of improved efficiency from which increased sales and profits and higher wages could be expected to flow. To the left-wing, Mondism was an extension of Havelock Wilson's policy to industry in general: a respectable Spencerism without the emotional overtones attached to the 'non-political' label. At the same time, the trade unions became non-political in a very real way: a new division being erected between political questions and industrial questions, the former to be solved by political means, and the latter by industrial means. The trade unions, as unions, were to concern themselves with industrial matters; and their political affiliation was not to affect their policy of co-operation with the Government of the day on non-political questions. This policy, which would have been quite un-acceptable in 1920 is nowadays rarely questioned.

Mondism was born at the 1927 Trade Union Congress when George Hicks in his Presidential Address, suggested an exchange of views between the two sides of industry to see: '. . . how far, and upon what terms, co-operation is possible in a common endeavour to improve the efficiency of industry and to raise the workers' standard of life.' This suggestion was taken up by an important group of industrialists and in January 1928 a joint committee was formed with the following membership: from the TUC, Ben Turner, Ernest Bevin, Walter Citrine, J. H. Thomas, Arthur Pugh, Tom Richards (who took the place of A. J. Cook who refused to serve) and Will Thorne; from the employers' side, Sir Alfred Mond, Lord Ashfield, Sir Hugo Hirst, Sir David Milne-Watson, Lord Weir, Lord Londonderry and plain Mr Vernon Willey. It was at first suggested that a National Industrial Council, representative of workers' and employers' organizations should be established and that it should operate a system of compulsory conciliation. However, this council was never established, but nevertheless, the principles underlying the con-clusions of the Mond-Turner talks were widely applied.

The Mondist policy of the TUC met with a great deal of opposition inside the Miners' Federation of Great Britain where acrimonious debates took place. The Secretary of the Federation, A. J. Cook, led the attack on Mondism by writing a pamphlet called 'The Mond Moonshine'. This drew from the TUC General Council the following protest:

'That this General Council, having carefully examined the pamphlet "The Mond Moonshine", express the strongest condemnation of the action of Mr Cook in issuing the pamphlet. The Council repudiates the pamphlet in its entirety, emphatically declares that it is full of inaccuracy, misrepresentations and deliberate falsehoods, and is obviously written for the purpose of discrediting and damaging the prestige and authority of the General Council. The General Council decided to bring this matter to the notice of the Miners' Federation of Great Britain, as Mr Cook's action is also a violation of the understanding come to at Nottingham on the 2nd March, viz., "that personal recriminations both by members and officials of both sides shall cease at once". We further request the Miners' Federation to state clearly whether they support Mr Cook's action.'[2]

Mr Cook explained to the Executive of the MFGB which considered this protest on Thursday, May 17, 1928, that this pamphlet, which was an expression of his personal opinions, was already in process of publication before the Nottingham pact was entered into, and it was agreed that the TUC General Council should be informed accordingly. The Executive also dissociated itself from the issue of 'The Mond Moonshine', and confirmed that the Vice-President of the Federation (Thomas Richards) should continue to serve on the Mond-Turner Committee.

Unfortunately, Mr Cook's explanation of the circumstances under which 'The Mond Moonshine' came to be published was not accepted by the General Council who renewed their complaint both in correspondence and at a joint meeting. The Council also protested at Cook's action during the Anti-Spencer campaign in Nottinghamshire, in cancelling meetings at which he was advertised to speak at Annesley and Eastwood and in speaking instead from a Young Communist League platform in Hyde Park on May 1, 1928. The substance of Cook's speech on that occasion (an attack upon the Movement's right-wing leadership and its Mondist policy) also caused offence. The General Council's complaint led Arthur Horner to allege that the Council was indulging in blackmail. Horner said:

'The truce there outlined [i.e. the agreement between the MFGB and the TUC General Council to put an end to mutual recriminations] was a condition of the General Council assisting an affiliated body, and by means of the fact, as shown by a later complaint, that the price of having their assistance to fight for recognition in Nottinghamshire was that they should not be fought in the advocacy of Mondism. Blackmail, I call it. . . .'

Horner was speaking at the 1928 Annual Conference which opened at Llandudno on Tuesday, July 17th. At the same Conference, the Forest of Dean put down a motion calling upon the Trades Union Congress to condemn the 'Peace in Industry' talks being conducted by the Mond-Turner Committee. The Interim Report of this Committee was seen by the mover of this motion (Mr J. Williams) as 'an agreement berween the General Council and the Mond Group to prop and support the crumbling edifice of British capitalism'. Mr Williams went on:

> 'Such concessions as the recognition of the Trade Union Movement . . . are the sugar coating round the bitter pill of rationalization.'

The Forest of Dean motion was seconded by a Nottinghamshire delegate. Mr W. P. Richardson, Treasurer of the Federation then moved the 'Previous Question'; but unfortunately the President, Herbert Smith, did not know how to deal with it, and in consequence the debate continued even after the 'Previous Question' had been voted upon.

Arthur Horner in supporting the motion, equated Mondism with Spencerism. He said:

> 'By rationalizing industry you eliminate human labour in every possible form and glut the unemployed market with still further masses of men. Then we are told that these proposals will give us a status, will give us a place in the sun as unions. We could always get a place in the sun as Spencer has done if we are prepared to transform the Trade Unions from instruments of struggle against capitalism into units of production for capitalism.'

The General Secretary of the Federation also supported the motion in a closely-reasoned speech. He pointed out that the origin of the Mond-Turner talks, a speech by the then President of the TUC, Mr George Hicks[3] was purely an expression of a personal point of view; further, that the response—a letter from the Mond Group of industrialists—was again a personal one. When the TUC General Council had considered this letter Cook had suggested that the matter should be referred to the constituent unions, but he had been over-ruled. Cook had also suggested to the General Council that some of the Mond Group had supported the Non-Political Trade Union Movement (which was probably true) and that the Council should insist upon the recognition of the Nottinghamshire Miners' Association and the repeal of the Eight Hour Act and the 1927 Trade Union and Trade Disputes Act before agreeing to enter the talks. However, his views had not been accepted by the General

Council. Instead, the view had been expressed that the miners should 'segregate' their industrial activities from their political activity. Cook had therefore, refused to serve on the Mond-Turner Committee, his place being taken by Mr Thomas Richards of South Wales.

Unlike Cook, Herbert Smith was opposed to the Forest of Dean motion. Smith was a pugnacious, overbearing Yorkshireman, but on this occasion he was apparently justifiably annoyed. He referred to the EC Minute concerning the appointment of Thomas Richards to the Mond-Turner Committee. This Minute had been so worded (by Cook) as to make it appear that 'the acquiescence of the chairman in the appointment of Mr Richards to the Sub-Committee was a personal acquiescence only and did not commit the Committee' whereas said Smith, Richards's appointment was made by a majority decision. Smith went on to allege that the issuing of pamphlets by the left wing was the cause of the decline in the membership of the Federation, and that the pamphleteers were 'playing Spencer's part'.

When the Forest of Dean motion was put to the vote, it was lost by 309 votes to 192. Nottingham voted for the motion despite the opposition of Frank Varley.

During the Mondist debates, references were made to the Minority Movement. It will be remembered that the Minority Movement was a left-wing 'ginger group' including both Communists and non-Communists in its membership. It was, however, Communist-led and inspired, although because of the Mondist policy of the TUC General Council, many Socialists who would not normally have associated themselves closely with the Communist Party did so in this period.

The Minority Movement had its origin in the South Wales coalfield; it was, indeed, part and parcel of the continuing syndicalist tradition. In Nottinghamshire the Movement became of some importance with the departure of Spencer and his colleagues. Now that the right-wing had its own separate organization, Nottinghamshire joined the small group of districts which were notoriously left-wing.

At its meeting on Wednesday, April 13, 1927, the MFGB Executive Committee had before it a motion from Nottinghamshire which reads: 'That we recommend one National Mine Workers' Union, and that a Special Conference be called to consider the matter as early as possible.' This motion was referred to a sub-committee who subsequently issued a questionnaire in which districts were asked to give details of their organizational arrangements.

At the Annual Conference of the Federation, which opened at Southport on Monday, July 25, 1927, Mr W. Carter of Nottingham seconded a motion on this subject which had been moved by Mr J. Williams of the Forest of Dean. The motion reads:

'That the National Executive Committee take immediate steps for transforming the MFGB into a British Mine Workers' Union catering for all persons engaged in and about the mines, accepting uniform contributions, and providing common benefits.'

After a long discussion this motion was withdrawn in favour of a much milder resolution from Durham which merely required the Executive Committee to formulate and recommend to districts such amendments to the Federation's constitution as were 'essential to its progress and effective work'.

Conference then went on to debate a further motion from the Forest of Dean which called upon the MFGB to 'take the necessary steps for securing a world-wide Miners' International'. During the discussion Mr Joseph Jones of Yorkshire read a document allegedly issued by Bob Ellis on behalf of the National Organizing Secretary of the Miners' Minority Movement, 38 Great Ormond Street, London, requesting Group Secretaries of the Movement to initiate resolutions at Branch level in support of the motions on 'One National Union and the Anglo-Russian Miners' Committee'. This allegation was described by Mr Williams of the Forest of Dean as 'a red herring'—a peculiarly appropriate expression—but, despite the red herring, Conference adopted the resolution by 511 votes to 261. Once more Nottingham voted with the left-wing.

At the same conference, Nottingham also voted in favour of an unsuccessful Scottish motion which supported the idea of allowing the Communist Party to affiliate to the Labour Party, whilst at the 1928 Annual Conference, Nottinghamshire nominated Arthur Horner for the Vice-Presidency of the Federation. On the other hand, when the Minority Movement was criticized at this Annual Conference, Nottingham voted in favour of the motion of criticism which reads:

'This Federation Executive Committee, after reviewing the general position in the Scottish coalfield, places on record its strong condemnation of the Communists and Minority Movement and the tactics which have been adopted in the various coalfields, and particularly in Scotland. It pledges itself to render all possible help to the bona-fide Scottish Miners' Federation and all other districts which are carrying out the principles of the Miners' Federation, the Trades Union Congress, and the British Labour Party.'[4]

The only districts to vote against this motion were the Forest of Dean with two votes and South Derbyshire with six: even South

Wales voted in support of the Executive on this issue. However, after the conference a manifesto, in which the conference resolution was strongly criticized, was published by Arthur Horner and bore the signatures of A. J. Cook; S. O. Davies (of South Wales) and Harry Hicken (of Derbyshire) among others. At the Executive Committee meeting held on October 12, 1928, these three Executive members agreed to withdraw their signatures from the manifesto and 'await the findings of the MFGB on the matters commented on therein and the inquiries in Scotland'.

A. J. Cook's precise political affiliations at this time are extremely difficult to sort out. That he was close to the Communist Party cannot be in doubt; and yet his name was apparently misused by them. For example, at the Executive meeting held on March 22, 1928, it was revealed that the Communist Party had made use of Cook's name in connection with the advocacy of local strikes in Northumberland and Durham. Cook confirmed that 'he had given no authority to the people concerned to use his name and that he was therefore not in any way responsible for their action'. Mr Cook also stated that he was personally entirely opposed to sectional stoppages, and believed that such stoppages in the present circumstances of the coal trade would be utterly ineffective in improving the position of the men. (Whether Cook's name was used without his authority, or whether this is another example of his ability to change his mind with his company it would be difficult to say.)

To sum up one can say that in this post-General Strike period, the TUC came out strongly in favour of the 'identity-of-interests-between-master-and-man' idea. They pursued a policy of peace, and they insisted upon a clear division between political and industrial activities. In the industrial field, they were concerned with securing the maximum efficiency in industry to the mutual advantage of employers and employed. So clear was the division between political and industrial questions that it was possible in the early 1930s for the Trade Unions to move to the right whilst the Labour Party, as a reaction to the defection of Macdonald, Snowden and Thomas, moved sharply to the left.

The Mondist policy of the TUC was little different from the policy of Spencer who 'always made it a cardinal point of policy to show the men that by a joint effort with the Management material benefits will result to the worker'.[5] This policy was however, distasteful to the left-wing which tended to crystallize around the Communist-led Minority Movement.

In Nottinghamshire, where the right-wing had left the field, the Minority Movement had much support. However, many of the most active left-wingers were unable to continue their association with the

NMA for long since, being blacklisted by the coalowners, they had to find work outside the industry. This applies, for example, to people like E. J. Ley and Les Ellis (both of whom became full-time officials of the union after the formation of the NUM), Owen Ford, German Abbott and a host of others. A few, like Bernard Taylor were able to maintain their connection with the NMA by being employed part-time as dues collectors whilst a few more, like Herbert Booth and Jesse Farmilo, were able to carry on as checkweighmen. There can be no doubt however, that the lack of vigour which the NMA was to display in the period 1930–1935 was due in part to the loss of its most active spirits.

CHAPTER 13

THE UNION'S LEADERSHIP

The Nottinghamshire Miners' Association had its home in the Leen Valley. All the original branches were in or near the Leen Valley; and the Union's headquarters was at Basford. The Erewash Valley pits were tardy in joining the Association and did not become properly organized until the 1890s. Even then, for a very long time the Leen Valley branches determined the Association's policy.

The Leen Valley pits were much more profitable than those of the Erewash Valley, and they paid better wages. There can be little doubt that the Erewash Valley branches profited by being attached to the more prosperous Leen Valley; since they enjoyed better terms of employment than those Erewash Valley pits which belonged to the Derbyshire Miners' Association. At the same time there was a very strong feeling among Leen Valley men that their wages were pulled down by the need to achieve some sort of parity between the two districts. Very naturally, a certain amount of friction was felt in the organization from time to time.

At the turn of the century the Leen Valley Branches held frequent meetings at the YMCA, Hucknall, at which they hammered out a joint policy on certain issues. Whilst these meetings were ostensibly held to decide on matters under negotiation between the union and the Leen Valley owners (The Nottingham and Erewash Valley Colliery Owners Association) in practice they also decided on nominations for the Presidency and Vice-Presidency of the Association—posts which were filled annually. During the period 1899–1906 the President was John E. Whyatt of Hucknall (and later, of Gedling) the Leen Valley's nominee; but in 1907–8 a nominee of the Erewash Valley, Charles Bunfield of Kirkby defeated him. In 1908, when a fourth full-time post was created, Charles Bunfield was elected to it, and his place as President was taken by William Carter of Mansfield.

The union was, at this time, led by individuals who were sincere no doubt, but colourless. The Agent, J. G. Hancock was an excellent administrator but a poor leader; the Secretary, Aaron Stewart was a tired man, much past his best; whilst the Treasurer, Lewis Spencer was and always had been, a nonentity—(He took the job in 1889 at a time when no one else wanted it.) Charles Bunfield and William Carter, who became a full-time official on the death of Stewart in

237

1910, were of a higher calibre, but the former at least was never accepted fully by the Leen Valley men.

George Spencer, who was elected President in 1912 (after a final spell in office by J. E. Whyatt), was a much more energetic and effective leader. Indeed, whilst J. G. Hancock nominally retained the leadership of the Association in practice George Spencer assumed it even though his was only a spare-time job. Spencer threw himself into the task of bringing the Erewash Valley wages more into line with those of the Leen Valley, and he campaigned also for a fair deal for the day wage men. Herbert Booth says[1] that Spencer's 'drive for better conditions for all won him general support; so much so that as time went on the Union called upon him to make out and advocate their cause in all the arbitration cases mostly with the Coalowners and also before independent arbitrators'; whilst Emmanuel Shinwell refers[2] to Spencer as 'a very active person (who) often went beyond others in demanding better conditions'.

In fairness to Hancock it should be borne in mind that he was a Member of Parliament from 1909 to 1922 so that he was away from the County a good deal. Further, he was temperamentally incapable of 'setting the Thames on fire'; his was a business-man's brain, an excellent man to run the office but very little use in the field. In the early days he and William Bailey had made a good team; now, a similar relationship developed between Hancock and Spencer. We have already seen that Spencer and Hancock worked together in 1914 to establish a political fund separate from that of the MFGB; and this partnership continued—with one short break—until 1937.

During the first World War, Frank Varley of Welbeck began to rival Spencer in the Association's counsels, and when Spencer was elected as a Permanent Official in 1918 Varley took his place as President. A year later Varley in his turn secured a full-time appointment. This appointment was made necessary by the increasing burden of work which accompanied the increasing membership and the sharpening of the conflict of interests in the industry. Even with a staff of six the Association was badly run in this period. Spencer and Carter followed Hancock into the House of Commons in 1918 so that, of the six supposedly full-time officials, three were available only at week-ends. Of the three who were left, only Varley showed any signs of life.

Carter and Hancock were defeated at the 1922 General Election which left only Spencer in the House, but in the following year Varley won Carter's old seat (Mansfield) so that now the Association's two most active officials were in the office only at week-ends and during recesses.

The five officials who acted as Agents worked in two rooms:

Spencer and Hancock in one; Varley, Bunfield and Carter in the other. Each of the five looked after one group of pits, and they appear to have kept fairly rigidly to their own pits so that as Herbert Booth says, during Parliamentary sessions, 'all hell was let loose in the Office between Friday night and Monday morning' whilst Spencer and Varley strove to catch up with their week's work.

This was a period during which the Association was faced with an enormous debt from the 1921 lockout, when the butties' organization was trying to push back the clock; when the left-wing Mansfield District Committee was seeking to commit the union to a Syndicalist policy, and when all too many miners refused to pay their subscriptions. Three of the leaders were inactive and the other two were so overworked as to be incapable of exercising effective leadership. If this period teaches anything, it teaches the unwisdom of allowing full-time Union leaders to enter Parliament.[3] Whilst Varley and Spencer were away from the coalfield, the Association was becoming more and more disunited; and the senior official, Hancock, was actually siding with the employers on some issues. We saw earlier that repeated attempts were made to dismiss Hancock but without success.

When A. J. Cook started to tour the County in the early 1920s he was held up by the militants of the left-wing as an example of what a Trade Union leader should be: incorruptible, energetic, ruthless in pressing the claims of his class against their exploiters. Cook was a leader of the Jack Lavin type: an advocate of one big union of mineworkers, of one universal brotherhood of man. The ground had already been well prepared for Cook's message in the Mansfield area. Any number of miners there were prepared to believe that Union leaders of the old type were corrupt and untrustworthy. Right from the beginning of the 1926 lockout, Cook made vague allegations of impending treachery on the part of people like Spencer and Varley: and these allegations certainly did not improve the atmosphere in those dark days. Indeed, Cook's week-end performances in the period preceding the stoppage and during the stoppage itself goaded Spencer almost to the limits of his patience. Varley, who was himself a wonderful platform orator, even if not quite so powerful a performer as Cook, was not 'nettled' to nearly the same extent. As Herbert Booth says, Varley was the only one of the NMA officials who could hold his own with Cook in the Mansfield district during the 1926 lockout.

Spencer and Varley knew Cook much better than the average Nottinghamshire miner did of course, and they were only too well aware of his fallibility. Herbert Booth who also knew Cook well (they were at the Central Labour College together and remained firm

friends until Cook's untimely death), considered Cook to be one of
the biggest liars he ever met and thought that he was capable of
'almost any kind of double-dealing'. This is perhaps, a harsh judgment and yet some very unsavoury episodes are recorded in the
MFGB Minute Book during Cook's term of office. We have already
seen how Cook was accused by left-wing delegates of talking their
way at open-air meetings, whilst making temperate speeches at
National Conferences, and how he was accused by Herbert Smith
of falsifying Executive Committee minutes in connection with the
appointment of Tom Richards to serve on the Mond-Turner Committee. We also saw that Cook's name was used widely by the Communists in support of their policies, and that Cook was wont to deny
any responsibility for the use of his name afterwards. He could not
however, deny responsibility for writing 'The Mond Moonshine'
(although he tried to wriggle out of the charge that he had published
it subsequent to the 'peace pact' made between the TUC and the
MFGB) nor could he deny having signed a statement published by
Arthur Horner in which the vote of censure on the Communists and
Minority Movement adopted by the MFGB Conference was attacked.
An even more unsavoury business was revealed by Mr Joseph Jones
of Yorkshire in 1928. He alleged that during the 1926 lockout, Cook
had negotiated with people who were close to the owners and the
Government without informing the Executive Committee of his
actions. This Cook strenuously denied, but a committee of inquiry
proved Jones to be right. On October 22nd the committee (F. B.
Varley and Ebenezer Edwards[4]) presented a lengthy report to the
Executive in which they say:

'From the evidence of Messrs Jones, Cook and Stuart (F. D.
Stuart, Private Secretary to Seebohm Rowntree) we have evolved
the following narrative: Where there was a conflict of evidence . . .
we have used the factor of "greatest probability" to reach a
decision. For example, when Mr Cook tells us that he had no
knowledge of any authority or backing possessed by these people,
we refuse to believe that the Secretary of a great organization
would sign a document such as this if drawn up by someone with
no standing. Again, when he tells us that he signed the document
but was not given a copy nor had seen one from that day to this,
we find it hard to believe that a responsible secretary to a million
men, fighting and starving for set principles, would sign a document agreeing to recommend "the working of a full seven hours
in Durham", "the extension of the shift work in Wales", etc., etc.,
things of which the men had not then heard. We cannot believe
that any sane man would let any such document out of his

7. W. Carter, MP

W. Askew

H. B. Taylor, MP

Frank B. Varley, MP

possession without retaining a copy of what he had agreed to as also of any reservation he had made and signed.'

The story in brief is this: Seebohm Rowntree and W. T. Layton (editor of *The Economist*) wished, in Layton's words, to 'formulate terms which the miners could accept and which the Government could present to the owners, and if need be, bring pressure on them to accept. Clearly it was of the first importance that an understanding between the miners and the Government should be reached before the Eight Hours Act was passed, and if this happened, that the Government should hold up the final stages of the Bill to enable a definite agreement to be reached'.

Rowntree, Layton and Stuart met Cook on a number of occasions (the first meeting with all three was on July 2, 1926) and formed the impression, as Rowntree noted, that '. . . he really wanted peace, and wanted moreover, a lasting peace and not a temporary one, and broadly speaking he was prepared to accept the [Samuel] report on condition that the Hours Bill was held up'. Sir Horace Wilson of the Ministry of Labour (who met Cook along with the other three gentlemen on the evening of July 2nd) was asked to try to delay the passing of the Eight Hours Bill conditional on Cook's signing a memorandum in which wage reductions were envisaged as well as a re-organization of the industry. Cook signed the memorandum as a basis for discussion without consulting Herbert Smith or any of his colleagues and handed it back to Stuart and Company on July 4th. During all this time Cook had been touring the coalfields with his famous slogan 'Not a penny off the pay, not a minute on the day'.

The committee concluded its report:

'1. That on July 3rd, 1926, Mr Cook did sign a document of which copies are submitted. (See Appendix to Chapter.)

'2. That when he did so, he knew that the said document would receive the consideration of the Government Departments.

'3. That Mr Cook was in possession of at least one copy (from which others could have been struck) which he could have brought before his EC at any time between July 6th and the holding of the Bishop's Conference.'

How on earth Cook could have entered into discussions in this fashion without consulting his colleagues is beyond comprehension. The strategy proposed by Rowntree and Layton was sound, and Cook is to be congratulated on seeing the soundness of it. It seems however, as though having signed the document he developed 'cold feet' at the thought of what the stubborn Herbert Smith, the suspect Spencer and Varley, and the militants on the left might say about his change of front. Cook was it seems, held captive by his own image. His militant youth—with its advocacy of revolutionary socialism, one big union for all mineworkers, and so on—had cast him for a Messianic role, and Cook found himself torn between the desire to play this role for all it was worth and the desire to work out an acceptable compromise. During the furtive negotiations with Rowntree and his colleagues, Cook continually evaded the making of decisions. In the report of the Committee of inquiry we have the picture of Stuart chasing about the country after Cook in an endeavour to obtain a decision of some kind. Thus:

> 'Stuart got hold of Cook after his meeting on Wrexham Race-course this same day, Monday, July 5th. After much persuasion, Stuart induced Cook to accompany him to the former's taxi to Cook's next meeting place, Hollywell. Stuart again urged him to come to London. Cook was nervously distraught, broke down and cried.'

Eventually Stuart prevailed upon Cook to return to London but when he went round to the MFGB Office to take Cook to meet Rown-tree and Layton he found that Cook had flown to the Continent.

Cook's actions were described by John Williams of the Forest of Dean as 'a piece of deception unprecedented in the history of Trade Unionism'; but this surely, is extravagant to say the least. Cook was an overworked man, a muddled man, a liar no doubt and a fool; but he was not a traitor. However, the reaction of Williams, who had been hitherto an extremist of the left and one of Cook's most ardent supporters, is understandable: his idol, it turned out, had feet of clay; his Messiah was, after all, a man like the rest of us. It is noticeable that from this time on John Williams sometimes opposed left-wing policies just as strongly as he had formerly supported them.[5]

Within a few years of the 1926 Dispute, there was a complete change in the full-time staff of the Association. Lewis Spencer had already gone (he died in 1927, but had earlier been demoted to care-taker). In 1927 also, the NMA found itself unable to support four officials and a ballot was therefore taken to determine which two officials should be asked to retire. Not unnaturally, it was decided that Bunfield and Hancock should be the ones to go. Immediately on relinquishing his employment with the NMA and after accepting a parting gift of £100, Hancock became Treasurer of Spencer's Union: an act which brought contumely on his head.

Two years later Frank Varley died, to be followed by William Carter in 1932. Val Coleman was elected to replace Varley and William Bayliss took Carter's place; while Herbert Booth, who had contested both elections, became President and William Askew became Vice-President. The senior official was now Val Coleman (he was called the General Secretary whilst Bayliss was called the Financial Secretary) and he was much the least able of the four. Booth and Bayliss on the other hand, did not see eye to eye on the only issue which really counted in this period: the question of unity. As we shall see in our next chapter, all too many of the leading members of the NMA failed to appreciate that their really important task was to bring back the Schismatics into the fold.

APPENDIX

Document signed by A. J. Cook in 1926.
(Copy)

Covering Letter to Document

I want to make a very strong appeal that the proposals which I have agreed to support shall be seriously considered as an alternative to the Hours Bill. I do not make the statement as a threat, but I am perfectly certain that if the Hours Bill goes through, it will create an insurmountable obstacle to negotiations. Even if the owners succeed, as they hope to do, in forcing the men back to work on the terms of the Eight Hours' Bill, they would create a situation which would mean, for many years to come, a bitter struggle to restore the seven hours. There will thus be a rallying point for agitation until that Bill is repealed, and there will be no rest and no goodwill in the industry. Moreover, the Miners' Federation will have the whole Trade Union Movement behind it in its agitation for the repeal of the Act. On the other hand, if the Government will accept the proposals herewith put forward as an alternative to the passing of the Act, *as a basis for discussion*, the Miners' Federation will do everything they possibly can to increase output and to develop machinery for securing permanent peace and goodwill. In a word, the alternatives for the coal trade are: 'Five years' peace, or five years' unrest'.

A. J. Cook.

July, 1926.

(Words in italics were underlined in ink by Mr Cook on the original.)

(Copy)

PROPOSED TERMS.

Confidential.

1. Conference to meet to arrive at agreed interpretation of report —including question of stage to which reorganization proposals

should be carried before any wage changes that may be required in conformity with the report become operative.

2. Conference to be in first instance between two sides and the Government. Failing agreement, interpretation to be left to Samuel (or Commission). Decisions as to interpretation to be reached within four weeks (or other agreed period).

3. Parties to agree in advance to put into effect the whole report as interpreted by the Conference.

4. Conference to endeavour to estimate financial results of re-organization, and the wages problem will be discussed in the light of these estimates.

5. The Conference will endeavour to reach an agreement, both as to the wages to be paid in the period following the time limit, and as to the permanent machinery for wage fixation.

6. Present subsistence rates not to be altered except by mutual agreement, or by recourse to the machinery already used for the purpose of fixing them.

7. If the Conference cannot agree during its deliberations as to the readjustment of wages to come into effect at the end of the time limit, the question to be remitted to a wages board, as contemplated by the report, containing a neutral element with power to determine the necessary readjustments as from the end of the time limit contemplated in paragraph 2.

(The permanent machinery of wage fixation must not be on the basis of compulsory arbitration, i.e. the men will not be asked to surrender the right to strike. But the machinery would not necessarily exclude voluntary arbitration. If, however, the machinery includes the right of rejecting arbitration in any particular controversy, the wages fixed in accordance with this paragraph must be valid for a period to be fixed in advance.)

8. Men to resume work forthwith on April terms, any losses during the period fixed in accordance with paragraph 2 to be made good, either by a loan secured as a first charge on the total proceeds of the industry, or by some subsequent readjustment of the proceeds of the industry (recoupment).

9. With a view to increasing output, and so reducing the cost of production, the Miners' Federation will undertake:

(a) To co-operate in a scheme to deal with voluntary absenteeism, if necessary by penalties.

(b) To ensure the working of the full shifts in Durham and Northumberland.

(c) To work double shifts in South Wales or any other area in which it is practicable and is desired by the owners.

(d) To discourage any restriction of output, and to co-operate in the establishment and full utilization of machinery to settle any question of alleged restriction.

(*e*) To assist in the extension to as many grades as possible of the principle of piecework, or some other system of payment by results.

I am prepared, speaking for myself, on condition that the Government does not proceed with their Hours Bill, to recommend my officials and Committee to consider these proposals as a basis for discussion.

A. J. COOK.

July 3rd, 1926.

(Words in italics were underlined in ink by Mr. Cook on the original.)

Certified to be a true copy of document signed by 'A. J. Cook', and submitted to us September 12th, 1928.

F. B. VARLEY.
E. EDWARDS.

7*a*. If, after taking into account the financial results of reorganization referred to in paragraph 4, the Wages Board finds itself compelled by the conditions prevailing in a particular district to fix wage rates other than those of men on the subsistence minimum substantially below the present minimum or alternatively at less than 25 per cent above standard, the Board may propose an alternative either no change or a smaller reduction in wage rates, combined with an increase of hours, provided that the alternative of increased hours shall only come into effect if on a ballot in the district concerned 66 per cent of the miners voting express their preference for that alternative.

(Cook did not sign this additional clause.)

CHAPTER 14

IN THE WILDERNESS

Inside the Nottinghamshire Miners' Association there were two schools of thought. One school, led by Herbert Booth, held that serious attempts should be made to re-unite with the Industrial Union, whilst the other, led by William Bayliss, believed that there should be no compromise with Spencer.

In the period 1926 to 1937 Booth toured the County with his advocacy of fusion of the two unions. In so doing, he created a certain amount of enmity. There were those for instance, who felt that it would be immoral to negotiate with Spencer. And there were others who had a vested interest in disunity, although doubtless many of these also were opposed to discussions with Spencer as a matter of principle. I refer to the Association's collectors.

These collectors held the Association together in its dark days. Some of them stood on street corners to collect subscriptions; others went round from door to door. In 1932 there were nineteen such collectors. Their takings ranged between £2 14s 5d per month and £31 18s 0d per month, and their wages between 2s 6d per week and £2 per week. Some of these collectors acted also as branch secretaries; and indeed, a few, where branches were not functioning, acted as Secretary, Delegate and Committee all rolled into one. Mr Herbert Booth feels that the decline in his popularity in this period was due to the hostility of some collectors who took objection to his unity propaganda.

However, Booth had an ally in A. J. Cook. In 1929 Cook was very concerned about the two serious breakaways: the Spencer breakaway in Nottinghamshire and the Communist (alias Minority Movement) breakaway in Scotland. The situation in Scotland was extremely involved, and all the fault did not lie on one side. Indeed, in Fifeshire the Communists had the majority of the men on their side for a time and the actions of their orthodox opponents leaves much to be desired. However that may be, and the details obviously lie outside the scope of this study, A. J. Cook subscribed to a report on the Scottish position (signed also by Smith and Richardson) which included this paragraph:

'We protest against the continued interference of the Communist Party (Minority Movement) and other bodies in the business of

246

the Miners' Federation of Great Britain and its affiliated organizations, and we call upon the members to resist this interference and the abuse of individuals which accompanies it.'

Cook also discussed this matter on a number of occasions with Herbert Booth. He told Booth that he was determined to bring both the left-wing breakaway in Scotland and the right-wing breakaway in Nottinghamshire back into the Federation.

At the meeting of the MFGB Executive held on September 12, 1929, it was reported that the officials had met representatives of the NMA 'with a view to obtaining assent for the Federation to proceed in their endeavours to obtain a satisfactory settlement in accordance with the circumstances prevailing'. Subsequently, Cook met Spencer privately through the good offices of the then Secretary for Mines, Ben Turner, who was present at the first meeting.[1] Apparently Spencer was not disposed to try to reach an accommodation with the NMA at first, but pressure was brought to bear on him by Cook and Emmanuel Shinwell. Unfortunately no record was kept of the many talks which took place; these were indeed, conducted under conditions of strict secrecy.[2]

Eventually on July 2, 1931, the four principal officials of the NMA (William Carter, General Secretary, Val Coleman, Financial Secretary, William Bayliss, President and Herbert Booth, Vice-President) were called to meet Emmanuel Shinwell (who had taken over the Mines Department from Ben Turner in 1930). Shinwell asked these four whether they were prepared to negotiate with Spencer with a view to achieving a fusion between the two unions. Coleman and Booth were in favour of meeting Spencer on July 27th but Carter and Bayliss were opposed to it. Subsequently, Council also turned down the idea and so it was not pursued any further.

Instead, the Association led an increasingly fruitless existence. The principal raison d'être of trade unions: negotiations with employers was denied to it; and even compensation cases had to be settled, in the main, through the Association's solicitors. Solicitors' bills therefore tended to absorb a disproportionate amount of the union's income. The Association had scored a minor success in 1928 when it applied to Thomas Hollis Walker, KC, acting for the Joint District Board set up under the 1912 Minimum Wage Act, for a revision of the legal minimum wage rates in the district. This claim succeeded, it being held that the application had the support of 'a considerable body of opinion amongst the workmen'. The actual benefit to the men was slight, if not non-existent, but the whole point behind the application was to force a measure of recognition for the Association. The author of this move was Frank Varley, whose wise counsels were soon to be lost to the Association. After his death the Association became increasingly sectarian in outlook; and too many of its leading members looked upon the absence of recognition as a proof of their own virtue. It is extremely improbable that this willing acceptance of martyrdom would have developed had Frank Varley lived.

The chief service rendered by the Association to its members in this period was the relief of distress during unemployment. Out-of-work pay absorbed an almost unbelievably high proportion of the union's income. Thus in 1932, out of a total income received at Head Office from contributions of £7390, no less than £4000 was handed back to members in out-of-work pay; and in the four years 1931–34 inclusive, out-of-work pay totalled £14,632. The Association also represented members very successfully at Courts of Refrees (which determined whether a person was entitled to unemployment pay from the State). The following extract from the Annual Report for 1934 will give some idea of the extent of the assistance given in this direction:

'COURT OF REFEREES.

'This department has now become a definite part of the objects of the NMA, although strictly political, it has a direct bearing upon the economic conditions of our members. It controls the incomes into the homes of men who are either on the PAC Means Test or Unemployment standard benefits. 54 cases have been represented during the six months. 41 have received direct benefits as a result. It is not possible to measure in cash the amount claimed, or even the number that will be entitled to benefit; but we have a record of which we are proud when we say that no less than 900 cases have been represented since 1929. Apart from establishing benefits on test cases, we have with success made several appeals to the Umpire, and even though this is a political wing attached to the NMA and huge benefits have come to the homes of the miners, many still listen to the advice of the employers to refrain from joining the NMA. It is not possible to give you the varied text on which these men must face the court of referees, in some cases men refuse to pay fines, some refuse to pay for timber, others refuse to work for wages less than district rates, some refuse to work overtime, and in each case these men prefer to face the Court Referees rather than be the forerunners of further cuts either in wages or in conditions.'

The advocates in these Courts of Referees were branch officials of the union, and they certainly deserve praise for the skill with which they presented their cases.

Because of the Association's inability to represent its members' interests in the ordinary affairs of working life, membership dwindled until 1935. The membership fell from 15,740 in 1928 to 13,475 in 1930; 13,315 in 1931; 12,295 in 1932; 9,985 in 1933 and 8,500 on January 1, 1935. To some extent this fall in membership was due to victimization. Many first-class miners were dismissed during this period for refusing to join the Industrial Union; and doubtless many more yielded to the pressure which was brought to bear upon them. However, this victimization is only part of the story. The real difficulty was a lack of resolution among the men.

Much the strongest branch was Kirkby, many of whose members were Welshmen. This branch had 2,000 members at the beginning of 1932 and 1,700 at the end of 1934. Apart from Kirkby, the strongest branches were in the old mining districts: Newstead (with 706 members at the end of 1934); New London (450 members); Clifton (444 members); and Bestwood (400 members). At the other end of the scale, many of the new mining villages had only a handful of adherents: Thoresby (with 3 members); Harworth (6 members) Ollerton (60 members, of whom 25 were 'financial'); and Bilsthorpe (50 members, of whom 40 were 'financial'). Gedling, where the butties were very powerful also had a very weak branch (35 members) whilst the Huthwaite and Welbeck branches had no members at all although there were fifty members working at New Hucknall Colliery. Both New Hucknall and Welbeck belonged to the New Hucknall Colliery Company; as did Annesley and Bentinck. The following circular issued by the Industrial Union at Annesley Colliery will give some idea of the pressure brought to bear on the men employed by this company.

'Annesley Colliery.
'21st December, 1931.

'Dear Sir,

'We are appealing to you to honour the agreement you entered into when you signed on at this colliery, i.e. to have your union contribution stopped through the office.

'We hereby enclose a form for you to sign if you still mean to honour your signature at the time you commenced at this colliery.

'Please hand back the form signed or unsigned to the person who hands you this letter, on or before the 29th December, 1931.

'Yours truly,

'B. Larwood.

(Secretary of the Annesley Branch of the Nottinghamshire Miners' Industrial Union.)'

'NOTTINGHAMSHIRE AND DISTRICT MINERS' INDUSTRIAL ASSOCIATION

'To the Manager—I hereby authorize and request you to deduct each week from the wages payable to me as a workman employed at the colliery, the sum of 6d., and to pay the same on my behalf to the Treasurer for the time being of the Nottinghamshire and District Miners' Industrial Union, and I declare that the receipt of such Treasurer for all sums so paid shall be your full and sufficient discharge therefore.

'I further declare that this request is made voluntarily by me and has not been imposed as a condition express or implied in or for my employment.

'Signed...,'

(Reprinted in MFGB *Minute Book*, 1935, pp. 10–11.)

Continual complaints of victimization were made against the New Hucknall Colliery Company with the result that all their collieries were badly organized. Even at Annesley, where Herbert Booth was the checkweighman, there were only 254 members, whilst at Bentinck, a very large colliery near Kirkby (and having, like Kirkby, many Welshmen on its books), membership was down to 395. This compares very unfavourably with Kirkby's 1,700 and represents only about a quarter of the potential membership.

Under these sorts of conditions, Labour people and Communists worked amicably together; with the result that the Association remained among the left-wing districts of the MFGB. Further, as the Fascist threat on the Continent grew, many of the Association's active members came to feel as Mr Bernard Taylor says, that their struggle was part of the fight against Fascism. And so we find Council adopting the following resolution in 1933:

'That in case of war being declared by this Government, we call upon the TUC through the MFGB to proclaim a general strike.'

Towards the end of the year, Council showed its concern at the events connected with the Reichstag Fire in a resolution which reads:

'That we protest against the new threat of death sentence being passed on the German and three Bulgarian prisoners who are being tried for the burning of the Reichstag, and we demand their immediate release; such resolution, if approved, to be sent to the *Daily Worker* and the German Embassy.'

This resolution was submitted by Kirkby. Another motion from the same branch which came before Council on March 24, 1934 reads:

'That the MFGB be instructed to collaborate with the TUC to organize a One Day Strike against the military oppression of the workers, and to consolidate the Working Class Movement of this country.'

In another resolution (submitted by Annesley) adopted in March 1936, Council agreed to support the application of the Communist Party for affiliation to the Labour Party. This resolution was to be expected. The Communist Party at this time was trying to build a United Front against Fascism, hence its application for affiliation to the Labour Party. In the Nottinghamshire coalfield, as we have seen an unofficial United Front already existed, and many of the members thought of themselves as partisans in the anti-Fascist War. (A few of them, indeed, like Bob Brown of Harworth, later fought with the International Brigade in Spain.)

This view was understandable though it had a regrettable effect upon human relations in the coalfield. Those who held it saw no possibility of compromise with the 'enemy': to them, Spencer was the lackey of the coalowners and a traitor to his class. The only way out of the difficulty was therefore to fight the issue out to a finish (although until 1936 the fight was an anaemic affair in all conscience). The propaganda carried out on behalf of the coalowners by the Economic League strengthened the militants in their resolve. The Economic League can be seen as the political expression of Spencerism. Its policy was a crude anti-Communism and anti-Socialism which did no good but which, on the contrary, added fuel to the fire of discord.

The Nottinghamshire coalowners were not fascists but they were undoubtedly shortsighted. It should have been obvious to them that they could not, alone of the district coalowners' Associations refuse to recognize the Miners' Federation of Great Britain for ever. Their refusal to treat with the MFGB may have been due, as they claimed, to loyalty to George Spencer; but it was a mistaken loyalty. It is very doubtful indeed whether the sponsorship of the Spencer Union brought the Nottinghamshire coalowners any real benefits. Spencer certainly looked after the interests of his members, and he used the

NMA as a bogeyman to frighten the owners, on occasion, into conceding claims which they had at first rejected. The solicitor to one
of the largest colliery companies in the county told the present writer
that they could never really trust Spencer. When an Agreement was
made with him it had to be fully spelt out or otherwise he would
examine every ambiguity in order to try to extract extra concessions
from the owners. Another representative of one of the old colliery
companies recalled an occasion when his company had been negotiating a new price list with Spencer. The negotiations were still at
an indeterminate stage when the local papers published a statement
by Spencer which implied that the owners had met the men's claims
fully. The company protested to Spencer that his statement was
untrue, but nevertheless they were unable to repudiate it publicly,
and Spencer refused to withdraw it. Indeed, he pointed out that unless
the published terms were adhered to, the NMA would be able to make
capital out of the situation and the men would come out on strike.
The company had no option but to pay.

In much the same way, Spencer fought compensation cases just as
energetically as did the NMA; and indeed he built for himself in this
period an enviable reputation as an expert on Compensation Law.

It was claimed for the Spencer Union that it had built up a system
of conciliation which saved the owners from troublesome strikes,
but even this is a doubtful claim. The chances are that the Nottinghamshire coalfield would have been virtually strike-free in this period
even had the breakaway not taken place. Spencer may have been an
insurance against further national disputes of the 1912, 1921 and 1926
variety, but the insurance policy lay idle. So far as local disputes were
concerned, the Nottinghamshire miner had had the fight knocked
out of him by the struggles of the '20s; and he was only too willing
to settle disputes peacefully. The very heavy unemployment in the
county was another factor which made strike action unlikely. The
Spencer Union was a device formed to suit the conditions of 1919–26
when the militants in the Miners' Federation had the upper hand. But
with the adoption of Mondism, the Trade Union Movement in general, and the Miners' Federation in particular, turned its back on
militancy; and the device of non-political unionism was rendered
obsolescent.

Towards the middle 1930s, far from damping down industrial
conflict, the Industrial Union was responsible for generating it. The
discontent in the coalfield, aggravated day by day and week by week,
had so built up that once the dispute came out into the open a bitter
conflict was likely to develop. If the discrimination against members
of the NMA succeeded in driving some out of it, it built up a reservoir
of resentment among the rest. This reservoir was to boil over in 1936.

In fairness to the owners it should be remembered that the NMA militants were asking not for fusion but for the total surrender of the Spencerites. The latter were hardly likely to commit economic suicide, and the owners were honour bound to support them. The sensible solution of a fusion of the two bodies on mutually acceptable terms was still being advocated by Herbert Booth but the majority of the Association's leading members were opposed to it. They preferred instead to go deeper and deeper into a cul-de-sac from which they would have difficulty in extricating themselves afterwards.

As we shall see in our next chapter, the solution when it came, was forced upon the combatants from national level; and the union's sojourn in the wilderness then ended. However, a great deal of bitterness was to be generated before this happy conclusion was arrived at.

APPENDIX

LEGAL MINIMUM RATES FIXED IN JUNE 1928 ON THE APPLICATION OF THE NOTTS. MINERS' ASSOCIATION

TABLE I			TABLE II	
General minimum rates of wages applying to all Top Hard mines except Gedling.			*Specified rates fixed for mines other than Top Hard including Gedling.*	
1. Contractors in abnormal stalls	10	9	10	3
2. Stallmen when unable from shortage of trams, rails or the like to earn a day's wage	10	3	9	9
3. Daymen, experienced, on the coal face	9	9	9	3
4. Daymen, others, under 20 years of age	7	0	6	9
5. Daymen, others, 20 years of age and over	8	6	8	0
6. Stonemen, Rippers, Getters out, and timbermen on contract	10	0	9	6
7. Datallers, chargemen	10	0	9	6
8. Datallers, others, under 20 years of age	7	0	6	9
9. Datallers, others, 20 years of age and over	8	6	8	0
10. Platelayers, head	9	6	9	0
11. Platelayers, others, under 20 years of age	7	0	6	6
12. Platelayers, others, 20 years of age and over	8	6	8	0

	TABLE I		TABLE II	
13. Corporals	8	6	8	0
14. Onsetters, chargemen	9	0	8	6
15. Onsetters, others	8	3	8	0
16. Horsekeepers, head	8	0	7	6
17. Horsekeepers, others	7	3	6	9
18. Haulage workers at 18 years of age	5	8	5	3
19. Haulage workers at 19 years of age	6	6	6	0
20. Haulage workers at 20 years of age	7	4	6	9
21. Haulage workers at 21 years of age and over	8	3	7	6
22. Haulage enginemen when wholly engaged below ground and recognized relief men when engaged as haulage enginemen below ground	8	6	8	0
23. Pump and boilermen when wholly engaged below ground at Mechanical Power Pumps and recognized relief men when so engaged below ground	8	6	8	0
24. Motor men when wholly in charge of and controlling motors below ground and relief men so engaged below ground	8	6	8	0
25. Air compressor and relief men when engaged below ground	8	0	7	6
26. Rope splicers	8	6	8	0
27. Coal cutter drivers	10	0	9	6
28. Jibbers and timberers	9	6	9	0
29. Cleaners out	8	9	8	3
30. Apprentices (grades 27, 28 and 29) at starting increasing by quarterly advances of 4d per day per quarter until the rate applicable to the grade to which the apprentice is drafted is reached	7	0	6	6
31. Boys, 14 years of age	3	0	3	0
32. Boys, 15 years of age	3	4	3	3
33. Boys, 16 years of age	3	10	3	8
34. Boys, 17 years of age	4	6	4	3
35. Boys, 18 years of age	5	2	4	10
36. Boys, 19 years of age	5	10	5	6
37. Boys, 20 years of age	6	8	6	4
38. Boys, 21 years of age	7	6	7	0

(Signed) Thomas Hollis Walker.

CHAPTER 15

HARWORTH AND AFTER

1. *The Harworth Dispute Develops*

The year 1935 saw some improvement in the Association's affairs. Membership during the year increased by about 1,500 (from 8,500 to about 10,000). There was a similar increase in income. The Sherwood Branch almost doubled its membership during the year, whilst at Gedling a reorganized branch of nearly 300 members (compared with thirty-five at the end of 1934) was leading an active existence. A new branch had been formed at Harworth, and by the year end it had 157 members—all financial—on its books. Booth and Bayliss were doing missionary work at Bilsthorpe and Ollerton with some small measure of success. Altogether the picture looked brighter than for some time past.

To some extent, the Association was benefiting from the effects of the improvement in the economic climate. The output of coal in Nottinghamshire, Derbyshire and Leicester rose from 29,165,790 tons in 1934 to 29,237,901 tons in 1935, and it was to rise still further —to 31,028,962 tons—in 1936 and 33,915,985 tons in 1937. Further, despite the increased membership out-of-work pay was down. In 1935 this cost the Association £2,323 compared with a yearly average of £3,658 for the previous four years. The upswing of 1935 was, of course, mild by any standards; but it did at least bring some hope of future improvement. At any rate the Miners' Federation decided that the effect of the upswing on the coal mining industry was sufficiently marked to warrant an advance in wages.

Accordingly, the MFGB sought the support of the Government for a flat rate addition to wages of 2s a day and for the re-establishment of National negotiating machinery; but the Government at first proved to be unhelpful though sympathetic. Not unnaturally, they took the line that they could not introduce legislation to secure an increase in wages for miners alone, nor could they force the owners to take part in National negotiations. The Secretary for Mines did, however, urge the Mining Association to negotiate with the MFGB but as usual he was told that the district coalowners' associations would not hear of such a thing.

The Miners' Federation on this occasion showed that it meant business. It organized meetings all over the country with the help of the TUC and Labour Members of Parliament; and it issued some thousands of leaflets. The Government proved sympathetic throughout, as one might expect during a General Election year. Under the circumstances, the attitude of the owners underwent a far-reaching modification; but not before a National Ballot had revealed a majority of 380,136 in favour of coming out on strike over the issue.

The actual offers received and accepted were well below the 2s a shift demanded, and they varied from one coalfield to another. Most of the Midland districts obtained an advance of 1s a day for men and 6d for boys; (but Derbyshire obtained only 9d and 4d respectively); whilst in some districts the increase was as low as 5d or 6d a shift for adults. The owners also agreed to the setting up of a 'Joint Standing Consultative Committee for the consideration of all questions of common interest and of general application to the industry, not excluding general principles applicable to the determination of wages by district agreements'.[1] The acceptance of this principle was of greater importance for the Federation than the size of the immediate increase.

In Nottinghamshire the wages campaign assumed a special significance. It was realized from the start that if a National Strike were to be called, the majority of Nottingham miners would continue to work. The campaign was therefore designed to achieve a substantial increase in membership, and was moderately successful as we saw above. A number of rallies were held (at which such speakers as Bill Betty of South Wales, Herbert Smith of Yorkshire and Joseph Jones, President of the MFGB appeared); the Red Flag was sung and broadsheets were issued by the thousand.

The Nottinghamshire coalowners conceded an increase of 1s a day for men and 6d for boys with effect from January 1, 1936. Credit for this was claimed by the Industrial Union; but the NMA declared in its Annual Report that '. . . it is pleasing to note that the miners in Nottinghamshire are fully aware that such a claim is groundless, they know full well that the present improvement in wages is entirely due to the efforts of the Miners' Federation and the Nottinghamshire Miners' Association'.

By the beginning of 1936, when the Mining Association agreed to the setting up of a Joint Standing Consultative Committee, it was obvious that the Nottinghamshire coalowners could not withhold recognition from the Federation for much longer. Already, during, 1935 there had been a great deal of discussion about this. The Executive Committee of the MFGB met the NMA Council at Nottingham on Saturday, February 2, 1935. Of the thirty-five delegates to Council, twenty were at work in the pits and the other fifteen (including collectors) were unemployed. Some of these delegates were understandably bitter; but it was pointed out by the MFGB representatives

HARWORTH MEN
Make History

Balloting at Harworth. Cross denotes Mr. Patchett, delegate

A STORY OF THE FIGHT
against Spencerism

Issued by
NOTTS COMMUNIST PARTY
22 St. Albans Terrace
Nottingham

PRICE ONE PENNY

8. Pamphlet issued by the Communist Party during the Harworth
dispute

that it was almost inevitable that a settlement could only follow negotiations with Spencer. In any case, the MFGB would not be prepared to 'resume responsibility unless it secured full power to complete a settlement'.[2] Accordingly, the following resolution was adopted by thirty-two votes to two with one abstention:

'That this Council meeting of the NMA agrees to the MFGB being given full power to take what action it desires to complete and restore this organization in its full capacity to negotiate for and on behalf of the miners of the Nottinghamshire coalfield.'

Subsequently, the MFGB tried to arrange a meeting with the Nottinghamshire owners, but they were told that 'the regulation of wages under the present (Spencer) Agreement was working satisfactorily and that they regarded it as impracticable to undertake negotiations with any other body.'

On the other hand, Evan Williams, the President of the Mining Association, made it clear that the owners' side would like the Nottinghamshire coalowners to belong to the Joint Standing Consultative Committee, which they clearly could not do so long as they gave sole recognition to the Spencer Union. The MFGB, as Joseph Jones its President stated, therefore 'took the opportunity of emphasizing the point that they are under an obligation to convene a meeting between ourselves and the Spencer Union in Nottinghamshire and therefore, we see the beginning of discussions with Spencer which in our view, will ultimately mean his elimination'. These discussions with Spencer did not however materialize for some time. Meanwhile the NMA suggested, in May 1936, that the men should be balloted once more 'on the question of the owners recognizing the Nottinghamshire Miners' Association'. The Executive Committee of the MFGB to whom this request was made, decided merely to 'make a suitable reply'. This left Nottinghamshire very much in the dark, so that at the Annual Conference held at Scarborough in July Herbert Booth asked the Executive what action they proposed taking to resolve the deadlock. Ebby Edwards, the General Secretary, said in reply:

'Nottinghamshire cannot be won unless the people themselves are prepared to fight. Let us be quite frank because some of us have been there weeks, and know the difficulty. Nottinghamshire holds the key position in the industry, wages are high, conditions and other things good, and yet they are one of the weakest links in the Federation.'

The fight, when it did start, developed into a rather more serious affair than Ebby Edwards—or anyone else—could have anticipated, although only one colliery was involved. This was Harworth, a newish colliery owned by the Barber Walker Company. Harworth is in North Nottinghamshire, quite close to the Yorkshire border; and the mining village built by the colliery company—Bircotes—acquired a character of its own. Physically, Harworth belongs to the South Yorkshire coalfield (and indeed, the Yorkshire Miners' Association had members employed there at times), whilst a high proportion of the labour force came from Durham. Bircotes was a raw place in the 1930s; a soulless village inhabited by people whose roots were elsewhere. Further, it was a company village: the houses were owned by the company; so was the land; the church, the parish hall, the Salvation Army Hut were all erected on land provided by the Company; and even the Curate-in-Charge looked to Barber and Walker for his £400 a year.[3] The atmosphere was quite unlike that of the old, settled mining communities of the Leen Valley. It was an atmosphere in which suspicion and mistrust bred all too easily. The radicals believed that their every action was reported upon to company officials, and this belief encouraged them to conduct their affairs in conspiratorial secrecy.

Harworth, despite the idyllic picture painted by the official 'historian' of the Barber, Walker Company, had grievances enough. The principal one concerned dirt deductions which were unprecedentedly heavy. The deduction per tram of coal was not less than 5 cwt. 7 lb. for two-fifths of the pit and not less than 4 cwt. 12 lb. for three-fifths of the pit. At meetings of the Industrial Union branch repeated attempts were made to have the dirt deduction agreement altered, but these attempts came to nothing since the Union maintained that, without the agreement, the pit would have to go on short time. (At this time, the maximum output of pits in Nottinghamshire and neighbouring districts was regulated by a Selling Scheme which laid down an output quota for each pit. Had the dirt deduction at Harworth not been made, the weight of the dirt would have been added to that of the coal for the purpose of arriving at the permitted output of the pit.) This argument did not satisfy the men however, since they held that the deductions for dirt were excessive and that they were therefore underpaid.

At the beginning of 1935 there were only seven members of the NMA working at Harworth out of a total labour force of 2,285. But during the year a profound change took place. During the month of June the membership increased from ten to 100, and during July a further forty applicants were accepted. At the September Council meeting, a hearty welcome was extended to the delegate from the new Harworth Branch. After a slight falling off in membership during the autumn, Harworth ended the year with 157 members. Six months later this had almost doubled (to 302) and there had been a complete change in the leadership of the Branch. Mr J. Pickering who had done so much to build up the Branch was succeeded as President by Mr Michael Kane, a Communist militant.

Shortly after this, the seething unrest at Harworth came to a head. The immediate cause of the dispute was simple enough in all conscience.[4] On August 31, 1936 Michael Kane and Ephraim Patchett, President and Delegate of the NMA Branch asked to see the Manager on behalf of the men (they wished to discuss among other things, an allegation that a deputy had struck two boys); but the Manager refused to see them since their union was not recognized by the Company. On the afternoon of September 1st, two men—who had been ordered by their deputy to 'stand snap' at 5.15 pm instead of 6 pm because of a haulage breakdown—refused to do so. They were instructed to see the Manager at the commencement of the following afternoon shift; but instead they succeeded in persuading 116 men to turn back from work. A deputation from the NMA attended upon the manager to discuss the grievance, but again he refused to see them. He also refused to discuss the matter in dispute with the representatives of the Industrial Union. The NMA branch therefore held a mass meeting and decided to post pickets at the pit gates in order to bring the men out over the issue of union recognition; with the result that only sixty-four men worked on the night shift; and only 125 worked on the day shift of September 3rd.

The Company then announced that the men were in breach of contract, and that they would need to re-sign before being allowed back to work. On September 3rd and 4th all those involved re-signed with the exception of twenty-five of the NMA's most active members who were refused employment.

Comparative peace then reigned until Thursday, September 17th when twenty men from 12s District came out of the pit, their complaint being that the ventilation was inadequate and that in consequence, their working places were excessively warm. Members of the management who carried out an examination rejected this complaint; and the Workmen's Inspectors, J. Prince and W. Moore (members of the Industrial Union) who examined the district on Saturday, September 19th also certified that conditions were normal, as they no doubt were by that time. Sixteen of the twenty men went to work as usual on September 18th.

'On September 24th the Manager received a report that four men on No. 11 conveyor face were calling others who were working there 'scabs'.[5]

'On Friday, September 25th, the Manager and Under-manager visited No. 11 face and told these men of the report. A man called

Wainwright admitted this and when asked why, he said he did not know. Another man, Wilson, was very insolent, and said he did not care what he did as he was single. They were told that they would be given seven days' notice as we could not allow this kind of intimidation from anyone.' The four men, Wainwright, Wilson, Winter and Turner received seven days' notice on September 26th when thirty of their workmates from 11s Face came out of the pit in sympathy with them. These men went to see the manager at about 10 am when they were told that, having left their work, they were regarded as having broken their contracts and that they were therefore no longer regarded as being in the company's service.

Meantime a strike ballot had been taken among the face men following the refusal of the management to re-employ the NMA militants who lost their jobs after the stoppage on September 2nd-3rd. This resulted in a vote in favour of striking of 785 to 136. For the moment this decision was not implemented; since as Joseph Jones informed the men at a meeting on October 11th, the MFGB were attempting to negotiate on the issue.

Instead, the face men decided to take action on the dirt deduction question. They held their checkweighmen, Clifford Parker and Smith responsible and they decided to get rid of them. A general meeting of contract workers was held on Sunday, October 18th, when a decision was taken to hold a ballot vote to determine whether the men had confidence in their checkweighers: the intention being that a vote of 'no confidence' would be tantamount to dismissal. Subsequently the ballot was held and a majority declared their lack of confidence in Parker and Smith who were therefore given notice by the Checkweigh Fund Committee. However, the notice was not valid for two reasons: the meeting at which the decision to take the ballot was made had been attended by people not in the Company's employ (the victimized NMA members) who had indeed taken an active part in the meeting although they were not entitled to do so; and the ballot was not on the question of whether the checkweighers should be dismissed. Counsel pointed out that 'A person may well have no confidence in an employee and yet continue such person in his employment'.[6]

Accordingly, when on November 16th the sub-weighman, Frank Holmes (who was also NMA Branch Treasurer) arrived at the weighbridge to take up his duties, he was informed by the management that they could not allow him to check the weights so long as the regular checkweighmen were still at work. That afternoon, the afternoon shift men had a meeting in the yard to discuss this matter, and all but eighty-five of them went back home. On the following day, Tuesday, November 17th only 195 colliers went to work.

Harworth Men Must Win!

Company Unionism must be crushed!

Since 1926, you in the Notts. Coalfield have been subjected to all forms of speed-up and tyranny, with victimisation rampant. Split and divided, you have been at the complete mercy of the Coal-Owners.

But you have reached the end of this period of help-lessness in face of attack. A new spirit is developing throughout the Coalfield. It is shown by the present actions of the men at Annesley, Bestwood, Linby—and above all by the challenge thrown down by the Harworth men to Barber, Walker & Co.

M.F.G.B. defied by Barber & Walker.

The tendering of Strike-notices at Harworth brings to a head the dispute which arose some months ago when the Management tried to impose worsened conditions on the men and boys underground. The Company have all the time refused to recognise the branch of the N.M.A., and after a Ballot—resulting in 1175 for the N.M.A. and 145 against the Company have sacked 800 men and now trying to make membership of the Spencer Union a condition of employment.

A Complete Victory for the Men at Harworth!

A complete victory is therefore of vital importance, not only to bring about better conditions at Harworth itself, but also to guarantee a successful campaign against Company Unionism, and worsening conditions throughout Notts. Finance is urgently needed to enable your workmates at Harworth to strike a blow for Freedom.

Give generously to the collection which will be taken at the pit gate on **Thursday.**

Printed by W. Mellors, Annesley Road, Hucknall, and issued by the Hucknall Labour and Communist Party Branches.

9. Leaflet issued by Hucknall Labour and Communist branches during Harworth dispute

Following their usual high-handed policy, the Company announced, on November 19th, that those men who were not at work were no longer in the Company's employ; and on the following day, as the Company admitted, 'instructions were given that men who signed on would sign a declaration that they were willing to become members of the Nottinghamshire Miners' Industrial Union'.

The Executive Committee of the MFGB met on November 19th and, after hearing a report from Herbert Booth, Val Coleman and William Bayliss, adopted the following motions:

'(a) That we advise the men to return to work and to deal with the question of the checkweighmen in a strictly constitutional way.

'(b) That providing they put themselves in a proper legal position, we empower the Federation officials to authorize the men to put in their notices at a date to be decided upon by the officials of the Federation.'

The MFGB had already, on November 4th, balloted the men to determine which union they wanted to represent them, the voting being: for the NMA, 1,175; for the Industrial Union, 145. It will be appreciated however, that the Spencer Union instructed its members not to vote so that the result is not quite so favourable to the NMA as it might seem at first sight. The fact remains though, that roughly a half of the Harworth men were willing to sacrifice their jobs rather than be dictated to in the matter of the union to which they should belong.

On Friday, December 11th the MFGB Executive meeting in Nottingham endorsed the advice given to the Harworth men by Joseph Jones and Herbert Buck (the EC deputation to Harworth) that those men still in work should hand in their notices. This brought the number of men living on strike pay provided by the Federation at the rate of £1 per week, plus 4s for each child up to 1,060.

The Barber, Walker Company stated that the number of men at work on any one day averaged around 600 for the three shifts. These men were escorted to work by the police. In an appendix to his report on the Harworth situation, Ronald Kidd, General Secretary of the National Council for Civil Liberties wrote:

'There have been a number of minor breaches of the peace and some scuffling with the police at the times when the men working in the Pit go to and from their work. These men are popularly called by the strikers "the Chain Gang". This procession of men forms up in a road some distance from the colliery entrance and is marched along under strong police protection. As soon as the procession has turned into the main Scrooby Road which runs past the colliery entrance, a police cordon is drawn across the road to prevent any crowds following and another cordon is thrown across the road farther on past the entrance.'

Most of the scuffling took place when the night shift went to work. The meetings of the men on strike were usually held so as to finish at the time when the 'chain gang' was on its way to the pit. With some hundreds of strikers pouring on to the streets at this particular time, the police were understandably harassed. This cannot however, fully account for the extreme measures taken by the police on occasion. For example, on one Sunday evening Mr Edward Dunn (who was then MP for Rother Valley) and Mr Bernard Taylor (now MP for Mansfield, and at that time a member of the Executive Committee of the NMA) were watching the night shift going to work when a young man standing peaceably near them was arrested for no reason at all. Mr Dunn went to the Police Station and informed the officer in charge that, unless the arrested man were released a full report on the matter would be made to the Home Secretary. This threat had its effect; the man being released.

Again Kidd reported that there had 'been many complaints that men and women proceeding to the Fish and Chip shop have been warned to get off the streets and have been turned back. From some parts of the Colliery Estate a very lengthy detour is necessary if the residents are to reach the shop by another way round. The direct approach is by the Scrooby Road. Complaints have also been made that persons going to keep appointments have not been allowed to pass the police cordon. On one occasion the fiancée of a young Harworth man had been spending the evening with him at his parents' home and they left the house just in time for her to catch the last bus back to Doncaster. The police refused to allow her to pass the cordon and she lost the bus.'

On the other hand, a witness (Mrs G. of Bircotes) said that 'On Tuesday evening, December 15, 1936,

'I was standing inside my garden gate arranging with two lady friends about going to Church on Wednesday evening when a policeman told me to "get inside". I said "I beg your pardon" and he replied "Never mind about begging anyone's pardon; get inside". I then told the policeman that my husband was working in the Pit and he apologized. I replied that I failed to see what my husband's working had to do with it.'

This statement is indicative of the favourable treatment received by the men at work and their families. This favourable treatment can hardly be wondered at however, since the main job of the police was to protect the persons and property of the strike-breakers and colliery owners.

A minority among the strikers certainly gave the police cause for vigilance. Stone-throwing and the use of catapults—particularly for window-breaking—were resorted to on a large scale; and attempts were made to hold up buses carrying strike-breakers to work.

2. *The Recognition Issue*

Meantime the MFGB Executive intensified its efforts to find a satisfactory settlement. As they saw it, the only issue which mattered was the recognition issue and this had to be pressed to a conclusion. Accordingly on December 30, 1936, Ebby Edwards wrote to Sir Alfred Faulkener, Under Secretary for Mines in the following terms:

'Dear Sir Alfred,

'Position at Harworth Colliery, Nottingham

'Further to my letter to you of October 7, I have to say that we have not even received an acknowledgment of my letter of September 30 to Messrs. Barber, Walker & Co. and in view of the complete failure to get a meeting with the manager or director of the colliery, my Executive Committee has decided to call a National Conference of the Federation for January 20 next to consider future policy in relation to the claim we make that the workmen at this colliery should have the elementary rights of citizenship in being permitted to join the Union of their choice.

'The dispute involves over 1,000 men, and I am afraid that a failure to settle the issue will have far-reaching consequences which it will be in the interests of all concerned to avoid. Hence this further request for your intervention, and I trust you will be able to use the good offices of the Department in endeavouring to effect a settlement before further and more serious trouble arises.'

Mr W. L. Cook of the Mines Department subsequently requested Spencer to ask the Company to withdraw the condition of employment that each workman must be a member of the Industrial Union. This Spencer did and he notified Cook on January 5th that the Company had agreed to the request, but that they were still not prepared to negotiate with the Nottinghamshire Miners' Association or the Miners' Federation of Great Britain. The MFGB refused to call off the strike on this basis.

Mr Cook met Spencer and Cooper again at their Clumber Street office on January 13th and told them that a national miners' strike would be inevitable if the recognition issue could not be resolved. According to a document prepared by the NMIU he was informed that

'. . . we were not going to be affected by statements of that character; so far as we were concerned we did not care a damn what they are saying and we were just as prepared for the fight now as at any other time; if it had to come, let it come now. He said he was there to avoid that.'

On January 20th a special conference of the MFGB met at the Friends House, Euston Road, London, to decide future policy on the Harworth Dispute. Mr J. E. Rowson of Lancashire suggested that consideration should be given to bringing all the Barber, Walker pits out on strike; but the Chairman, Mr Joseph Jones, pointed out that the Eastwood Collieries were 'strongholds of Spencerism and were unlikely to respond to a strike call.'

Mr Arthur Horner, on behalf of the Executive, moved:

'That this Conference of the Mineworkers Federation of Great Britain, having considered the position at Harworth Colliery, and the deliberate attempt of the organized employers in that district to prevent the mineworkers joining the organization of their choice empowers the Executive Committee to take a ballot vote of the entire coalfields upon the question of enforcing the principle of the freedom of organization and Trade Union recognition for those so organized.

'Further, it requests the Executive Committee to approach the General Council of the Trades Union Congress in order to enlist the support of the whole Trade Union movement for this principle, and in the meantime the Executive Committee to take all necessary steps to deal with questions which have been raised or may arise, and the Conference stands adjourned to be recalled by the Executive Committee.'

Horner said that this was not merely an organizational matter, but that:

'. . . we are trying to claim the monopoly of the sale of mining labour in this country: We are trying to get into the position where we can say to the owners, "You buy from us for you can buy from no other, and you must pay our price." Industrial unionism is an instrument in the hands of the owners to break the "corner" in mining labour which the Federation is trying to establish.'

Mr Joseph Jones pointed out that whilst the average selling price of coal in Nottinghamshire and Derby (at 13s 11·64d per ton) was slightly higher than in Yorkshire (13s 11·47d) the wages cost was $3\frac{1}{4}$d per ton less; and the profit was $3\frac{1}{2}$d per ton more; and he drew the conclusion that:

'what was saved in wages in the Nottinghamshire and Derby coalfield by Spencerism, went into the owners' pockets.'

After a long discussion Horner's motion was carried unanimously.

However, although the end to be achieved was clear enough, there was no unanimity as to the means which should be adopted to reach that end. The MFGB leaders saw a fusion with Spencer as the obvious solution; but a considerable body of opinion in the country wanted to see Spencer's organization crushed. This difference of view was most sharply defined in Nottinghamshire, where Herbert Booth the President, was for fusion whilst William Bayliss was for having no truck with Spencer. Val Coleman vacillated but tended to support Bayliss rather than Booth. The disagreement came to a head following a meeting at the Headquarters of the MFGB where Bayliss and Booth expressed their opposing views forcibly. Herbert Booth writes:

'We left the offices in Russell Square and for the first time in my life I discovered that it was possible to be in the company of men and be alone.'

At the next Council meeting, held on Saturday, January 31st, after reporting on the discussion with the National Officers, Booth resigned the Presidency of the Association, and left the Council Chamber.[7] This was the act of a tired man, weary with frustrated hopes. However, he was not to be given any peace. His home was invaded by journalists to whom he had to repeat his side of the argument over and over again. Even so, to the average member of the NMA, his resignation came as a shock.

Shortly afterwards, Booth was approached by Horace Cooper, Secretary of the Industrial Union and B. Larwood, Secretary of the Annesley Branch of that Union. By this time, it was fairly evident that fusion would result from the discussions going forward at national level, and it was suggested to Booth that if he would agree to join the Industrial Union, they would nominate him for office in the amalgamated body which would take the place of the two existing organizations. On this understanding Booth agreed to their proposal. This understandably aroused a great deal of bitterness among Booth's old associates, who characterized his act as treachery. Thus, William Bayliss speaking at a special conference of the MFGB on April 1st said:

'You will know we have one of our officials who has ratted during these troubles, and our [i.e. the MFGB] officials have indicated very clearly that the other that is referred to as representing the third official of the Industrial Union [i.e. the third nominee of the NMIU in addition to Spencer and Cooper to hold office in the new union expected to result from fusion] is the man who has ratted from us. I suggest to this Conference that they will not subscribe to rewarding treachery of that kind. I am hoping they will not. I feel convinced they will not.'

3. *Fusion at Last*

The Mines Department found their task of reaching an amicable settlement extraordinarily difficult. To start with, Spencer played 'hard to get'. Although he desired fusion as much as anyone he was determined to put the Federation in the position of supplicants. When he agreed to meet Federation officials he put on record that he was only doing so 'In deference to a request from the Secretary for Mines'; and he usually found it impossible to attend meetings on the dates suggested by the other parties. On the other side, the Federation officials could not make up their minds whether to insist on direct recognition, or to settle for fusion with Spencer. The rank-and-file were, in the main, opposed to any arrangement with Spencer, and this feeling proved an embarrassment to their negotiators. Spencer was not slow to exploit his opponents' differences. Thus, writing to the editor of *The Times* on February 11, 1937, he referred to a report of a speech by Mr J. McGurk, a member of the MFGB Executive Committee in which the following passage appears:

> 'Spencerism has been smashed at Bedworth [sic], and it will be smashed here (i.e. Harworth), and we shall have a united front at least of the miners. We will smash this tyranny, and I pledge myself to do everything possible to win support for a speedy victory.'

Spencer's letter continued:

> 'These words were used in a speech at Harworth last Sunday, and if they are a reflex of the policy of the Federation it appears they are offering one hand of friendship and in the other hold a weapon of violence, which they declare they intend to use against us.
>
> 'How is it possible with such a spirit prevailing to enter into any form of negotiation? We have been willing to listen to the voice of reason and to co-operate if there is a real spirit which would make for peace and prosperity in the industry, but we will not be intimidated by violent language into abandoning our rights to organize.'

The Nottinghamshire coalowners who were also brought into the discussions, made it clear that they supported Spencer up to the hilt. Indeed, one of the most influential of their number, Captain Muschamp is reported as saying in a speech at the Black Boy Hotel, Nottingham, on March 20, 1937, quoted by A. Horner at an MFGB Conference:

'We want to adopt the German idea. If the Government is to check future trouble it must put its foot down and put it down strongly. In this district we have been very much blessed with peace for the last ten years. This district—the Nottinghamshire district—can take credit to itself for having smashed the national strike, and since then, we have carried on peaceably with the Industrial Union for ten years. The country may thank the Industrial Union for preventing a strike a year last Christmas. I don't think that can be denied by anyone. This Industrial Union—I am speaking on it because it is a pressing point at the present time —has been the buffer which has prevented national strikes in this country. I am sorry to say and I think you will agree, that negotiations are now taking place which may have the result of smashing up the Industrial Union. We hope not. At any rate, the Mineworkers' Federation have tried very hard by frontal attacks to smash it. They have entirely failed. They have gone round the corner and with the assistance of the Mines Department, I am sorry to say, they are now trying to get in at the back door. Whether the Government thinks its policy of supporting the Mineworkers' Federation in doing away with the Industrial Union is for the good of the country or not, I don't know, but they are making a mistake because they are making possible national strikes, which would be the worst thing that could happen to it. There will probably be, as trade improves, dissatisfaction and strikes here and there. The working men of England do not want to strike. Generally speaking, they only want to be left alone.'

With support of this kind Spencer felt confident that fusion, when it came, would be fusion on his terms.

At the very first meeting with the officials of the MFGB held on February 25th, under the chairmanship of the Secretary for Mines, Spencer insisted that the following main principles should be accepted at the outset:

1. That the Agreement with the Nottinghamshire owners which was due to expire in March 1938 should be renewed (though with any amendments which might be mutually agreed upon) for a further five years.

2. That Nottinghamshire should be immune from strikes during the currency of the Agreement mentioned above (i.e. until March 1943).

3. That all local officials and collectors of the NMA who were not employed in the industry, should be excluded from membership.

4. That the Industrial Union should be allowed to nominate three full-time officials to the NMA's two.

5. That the rules of the amalgamated body should provide that political business could only be discussed at specially convened political meetings.

At a meeting of the Executive of the Federation held on March 4th, it was agreed by thirteen votes to six that the officials should accept these five principles as a basis for further negotiation. The negotiators then proceeded to the Mines Department where a further meeting with Spencer was to be held. At this meeting Spencer asked whether the Industrial Union could affiliate to the MFGB without amalgamating with the NMA. He was told that this was impossible.

Next, Spencer produced his five principles again, together with one or two more. These were embodied in a document styled the 'Draft Heads of Agreement' which Spencer and Joseph Jones signed. The officials of the MFGB had intended to go on to discuss terms for a settlement at Harworth, but Spencer left the meeting with the draft 'Heads of Agreement' in his pocket, and the Harworth question as much in the air as ever. Ebby Edwards described what followed next in these words:

'After he left, we were alone with the officials of the Mines Department, and the Minister was—I think—as much perturbed as we were. There were the headings set out. Harworth was still standing as a dispute. As already indicated Harworth was linked to the terms. I need only explain this—that the Minister agreed immediately that the Industrial Officer, Mr Cook, should go off to Nottinghamshire, he should get Spencer and also travel to the Harworth Colliery, in view of the circumstances, to try and get some conditions for the orderly return of the men and the prevention of victimization.'

Subsequently, Mr W. L. Cook met the officials of the Barber Walker Company together with Mr Spencer. Following this meeting this report was issued:

'This meeting took place at the Harworth Colliery on March 5, 1937. There were present: Messrs D. MacGregor and W. Wright. Messrs W. L. Cook, for the Mines Department, and G. Spencer.

'1. The management cannot meet any representative of the Mineworkers' Federation until amalgamation of existing unions is an accomplished fact, and publicly announced as such.

'2. The Management are prepared for new men now idle to resume work at the rate of thirty to fifty per week, gradually increasing until a maximum of 350 are employed.

'3. The Management will select from available men those who, in their opinion, are most suitable for the various jobs now requiring to be filled.

'4. Men already at work are feeling bitter about working in future with the men who are now unemployed, but the Management, with representatives of the Industrial Union, are prepared to meet the men at present working, for the purpose of persuading them to establish good relations with those who will be in future employed.'

This report was considered by the MFGB Executive at its meeting on March 12th. The Committee also had before it a telegram from David Buckley, who was now Secretary of the Harworth Branch of the NMA asking that '. . . before any agreement with Spencer is signed, the whole question be referred back to the districts re ballot of the whole coalfield'.

The Committee meeting adjourned in order that the Officials might see Captain Harry Crookshank, Secretary for Mines, on this issue. Captain Crookshank met the officials at 3.45 pm and expressed his willingness to try to secure from the Barber Walker Company '. . . proper assurances that there would be no victimization at Harworth'.

On March 24th, Ebby Edwards met the Secretary for Mines again. Captain Crookshank had met the owners with Spencer on the previous day, and the owners had undertaken to reconsider their stand on the Harworth Dispute. While Edwards was with the Minister, a telephone call was received from Mr F. H. Ellis of the Nottinghamshire Coalowners' Association in these terms:

'1. Following upon the suggestion made by the Secretary for Mines at the meeting yesterday, the management of Harworth Colliery agree that, if and when amalgamation of the two Unions is an accomplished fact, they will meet Mr Spencer with Mr Coleman, or any two permanent officials of the new union whom they may respectively choose as their representatives, to discuss any grievance that may arise out of the re-employment of men not now working at the colliery.

'2. That no misunderstanding may arise, the Company again declare that a change in the shift now and hereafter to be worked at the colliery, makes it impossible to re-employ more than a maximum of 350 men. These will be employed as vacancies arise.

'3. These paragraphs are supplemental to the statement made by Mr McGregor to Mr W. L. Cook on March 5, 1937.'

The Executive Committee of the Federation, meeting later on the same day, decided to call a special conference to formulate policy for Thursday, April 1st. This conference duly met at the Kingsway Hall, London.

The Issues in the Nottinghamshire Dispute

A Message to the Miners of Britain:

In the dispute in the Nottinghamshire coalfield certain principles are at stake of outstanding importance to every mineworker in the country. The first is the simple, but vital, issue—shall mineworkers who desire to join and be represented by their own Federation be free to do so? The second—shall men be victimised because they have fought for their rights as Trade Unionists? If a definite answer in the affirmative cannot be given to the first question and an equally emphatic "no" to the second, then we are no longer free citizens, and Trade Unionism has lost its meaning.

You will know that in the Nottinghamshire coalfield a satisfactory answer CANNOT be given to these questions. There, large numbers of men have been compelled to join a Union, not of their own choice, but of their employers' choice, and the heroic men of Harworth pit, who, for five months, have fought the battle of free Trade Unionism, are in grave danger of victimisation.

Deeply involved in this dispute is a further and even more vital issue—the future welfare of the whole of the mineworkers of the country. So long as the employers in Nottinghamshire are able to divide our forces, so long shall we have a fatal weakness and be unable to bargain with our full powers, and so achieve peace and justice in the coalfields. Great as are the other issues involved, this overshadows them all; for, in this matter, the whole future of the mineworkers is at stake.

Not wishing to embroil all the coalfields, and desiring, if at all possible, a peaceful solution of the difficulties, the Federation has sought to achieve this by way of merging the two forces in the Nottinghamshire coalfield, and so eventually to create a better and healthier situation. But the present terms of the Industrial Union are not satisfactory; they are not terms which can be accepted by the great free body of mineworkers in this country.

And so you are asked to make a decision. By this ballot vote you are asked to record, not merely your detestation of the violation of the principles of free citizenship and free Trade Unionism in Nottinghamshire, but your determination to fight for those principles if necessary, and your Federation unhesitatingly recommends you to do this by recording an emphatic "YES" to the question asked you. Armed with your authority in this way, the Federation will then make further efforts to achieve a peaceful and honourable solution of the dispute, but every man must clearly understand that if, after being armed with this authority, the Federation is still met with a stubborn and vindictive attitude on the part of the employers, then it may be necessary to close the ranks and fight bitterly and stubbornly for those rights which are of the very essence of our liberties.

On behalf of the Mineworkers' Federation of Great Britain,

JOSEPH JONES, *President.*
WILLIAM LAWTHER, *Vice-President.*
EBBY EDWARDS, *Secretary.*

50 Russell Square,
London, W.C.
April 4, 1937.

---

10.   Leaflet issued by the MFGB in connection with the 1937 ballot

Notts. Miners' Association

To every Miner in Notts.

The M.F.G.B. is asking every miner in the British Coalfield to vote in favour of LIBERTY for you here in Notts.

You MUST Vote also for your own Freedom

The Spencer Union has never obtained One Penny increase in your wages—it has no power to do so

You MUST Vote to get that Power

Look out for the issue of Ballot Papers. See that you get one and mark it for the Miners' Federation

UNITY IS STRENGTH !

NOTTINGHAM PRINTERS LTD.

12. Leaflet issued by the NMA in connection with the 1937 ballot

11. Leaflet issued in connection with the 1937 ballot advising men to vote in favour of national strike action to obtain recognition of the MFGB.

A number of delegates were highly critical of the Executive for accepting the 'Heads of Agreement'. It also transpired that the NMA Council had reluctantly endorsed the national Executive's policy at a meeting on March 13th but reversed their decision, after consulting their branches, a week later. Val Coleman put the blame for this change of front on the Spencerites who had taunted NMA men with such terms as 'throwing in the towel'. Coleman went on to say 'Very good, if it is we have to die and never return, I think it would be infinitely better for Nottinghamshire to continue to remain on the sands rather than the Miners' Federation lend themselves to be dictated to and dominated over by Spencer.' Coleman was particularly incensed at the suggestion that Spencer should be President of the new Union for ten years where he would 'regulate, dominate, and finally carry into effect a policy which will be far reaching in the days that have to come'.

Ebby Edwards pointed out that a very difficult position would follow if the MFGB accepted the terms for fusion contained in the 'Heads of Agreement' whilst the NMA rejected the terms. [Spencer was claiming that, under these circumstances the MFGB would have to accept his union into affiliation, and disaffiliate the NMA. He appears to have been supported in this by Sir Alfred Faulkner, the Under-Secretary for Mines at a meeting on March 4, 1937 according to typewritten notes prepared by the NMIU.]

The sense of the Conference was clearly against fusion on the terms contained in the 'Heads of Agreement', and eventually the following resolution was adopted by 503 votes to thirty-two:

'That this Conference regards the draft terms for the formation of one Miners' Union for the Nottinghamshire coalfield as unreasonable and unacceptable. It also deplores the absence of any satisfactory assurance regarding the reinstatement of the men at the Harworth Colliery. It resolves, therefore, that a ballot of the coalfields be taken with the object of securing recognition of the Mineworkers' Federation of Great Britain and adequate assurances to prevent victimization at Harworth Colliery.'

The ballot vote was duly taken and it resulted in a large majority in favour of a national strike. What followed assumed almost the character of a farce. The Secretary for Mines and his staff met the MFGB and Spencer together with his allies, the Nottinghamshire owners, separately. The owners refused to give direct recognition to the Federation, and they refused to meet its representatives unless and until amalgamation of the two unions was accomplished. For his part, Spencer would not meet the Federation officials unless they were 'empowered to reach a final and binding settlement'.

S

In an endeavour to break the deadlock, the Executive of the MFGB next suggested that discussions should take place between representatives of the NMA and the Industrial Union. Unfortunately, the Federation Conference rejected the suggestion and 'decided that notices should be handed in to expire simultaneously throughout the Federation on May 22nd'.

On May 6th, the Executive of the MFGB met the Secretary for Mines and agreed to suspend the handing in of notices for a fortnight '. . . in response to the Prime Minister's appeal and in the hope of securing a settlement of the dispute honourable to all parties . . . providing an early conference of the three parties—the Nottinghamshire colliery owners, the Mineworkers' Federation of Great Britain and the Nottinghamshire Industrial Union can be arranged'.

At the invitation of the Secretary for Mines, the Nottinghamshire Owners and Spencer agreed to meet the Federation representatives but only on condition that they accepted the principle of amalgamation and that they would agree 'to submit to arbitration any points on fusion which after discussion could not be mutually agreed upon'.

This offer was rejected by the Executive who insisted that any meeting with the other parties must be unconditional. Arrangements were therefore made to bring the Federation's members out on strike on May 29th. The strike was however, averted. A face-saving formula which permitted negotiations between the parties to open in earnest was adopted and draft terms of agreement were laid down. These draft terms were ratified by a special conference on May 28th where it was resolved:

> 'That this Conference ratifies the provisional terms negotiated by the Sub-Committee of the MFGB with the Nottinghamshire coalowners and the Nottingham and District Industrial Union, under the neutral Chairmanship of Mr John Forster, and resolves that the notices in all districts, whether handed in collectively or by the individual workmen, be withdrawn.
>
> 'Further, the officials be empowered to render every assistance in the application of the terms and take such steps as are necessary with the Nottinghamshire Miners' Association to give full effect to the amalgamation.'

This settlement would not have been achieved had the officials of the Federation carried out Conference decisions. After all the harsh words spoken by Val Coleman and Bill Bayliss; McGurk and Horner; fusion was achieved substantially on Spencer's terms. Mr Frank Collindridge an EC member from Yorkshire admitted frankly: 'In the teeth of that ballot vote we pursued in our organization . . . a policy which in effect put that ballot vote at defiance.' The settlement provided that Spencer himself was to be permanent President, Herbert Booth was to be a full-time Agent, and the Wages Agreement with the Owners was to run to December 31, 1943. The MFGB had even, to use Mr Collindridge's words, given a 'moral pledge' that Nottinghamshire should be strike-free for five years.

As Coleman had predicted, Spencer held the reins in the new union. His was the deciding voice on policy issues. Further, to use the words of Herbert Booth in a letter to the author: 'After the amalgamation —free from Parliamentary worries—Spencer became the fighting leader again and carried on as such till his retirement in 1945. The present Old Age Pension scheme in Nottinghamshire is absolutely and entirely the work of Spencer.'

At the Annual Conference of the MFGB which opened at Blackpool on July 19, 1937, a resolution protesting at prison sentences 'imposed upon our Harworth comrades, and especially one defendant, who was a miner's wife' was passed unanimously. Among those voting for the resolution were George Spencer and Horace Cooper who were introduced by the President, Joseph Jones, in these terms:

> 'May I now remind the delegates in keeping with our desire at least to adopt the spirit of the new arrangement in Nottingham-shire, even if we cannot carry it out to the letter, that there is present at this Conference Mr George Spencer and Mr W. Cooper, of the Nottinghamshire Industrial Union. Some of you will re-member making a promise, and in keeping with that promise Mr Coleman and Mr Bayliss were invited to the next meeting of the Nottinghamshire Wages Board, and they (the NMA) have included Messrs Spencer and Cooper in their representatives to give them an opportunity to again meet the representatives of the districts in this Conference. In the name of this Federation therefore, I extend to them a cordial welcome and express the hope that they will continue their co-operation with us to the full to ensure that the remaining parts of the agreement are fully carried out at Harworth.'

Spencer replied: 'I thank you very much,—on behalf of Nottingham-shire.' He was well satisfied. Eleven years earlier he had been ordered out of a National Conference by a Yorkshireman; and now it was a Yorkshireman who welcomed him back.

4. *The Nottinghamshire Miners' Federated Union*

The Nottinghamshire and District Miners' Federated Union came into existence officially on September 1, 1937.

At the first Council meeting of the new body, it was reported that a ballot of the members of the two constituent unions had been taken, and had resulted in a statutory majority in favour of amalgamation on the terms provisionally agreed on May 26th. Mr E. Hopkin, the Union's solicitor confirmed that all legal formalities had been complied with, and that the Rules of the Federated Union (which he had helped to draft) had been registered by the Chief Registrar of Friendly Societies.

Service Agreements specifying the terms of employment of the five permanent officials (G. A. Spencer, President; Val Coleman, General Secretary; H. W. Cooper, Financial Secretary and Treasurer: H. W. Booth, Agent and W. Bayliss, Agent) were signed. It was agreed that G. A. Spencer should represent the Union on the Executive Committee of the MFGB until the Annual Conference of that body; and it was further agreed that the Presidents of the two constituent unions (Bernard Taylor, NMA, and Ben Smith, NMIU) should be Vice-Presidents of the new Union for a period of two years.

The following delegates were appointed Executive Committee members to hold office until December 31, 1939:

Ex-NMA:	*Ex*-NMIU:
Bert Parr	William Evans
Walter Harley	Joseph Birkin
Arthur Green	Tom Willoughby
George Syson	W. H. Hodgman
Clarice Alexander	N. Buxton
Ephraim Patchett	H. Willett
A. Dawes	One other to be appointed.

The branch officials were similarly appointed from among the old officials of the two constituent unions.

By the time the new union came into existence, the problems of Harworth had been substantially resolved. The labour force at Harworth had been permanently reduced during the dispute, and work was only available for some 350 of the 900-odd men who had been out on strike. Many had however, left the district or had found employment at other local collieries. Those still out of work were able to draw unemployment pay. Indeed, even during the dispute the Court of Referees had ruled that the men were entitled to unemployment pay since their employers were insisting on an unreasonable condition of employment—membership of the Spencer Union.[8]

A few of the men whose names had been drawn 'out of the hat' for re-employment found difficulty in taking up their jobs owing to the refusal of ex-Industrial Union men to work with them. However, Spencer did his best to make his members honour the agreement entered into with the employers. The dispute had left its scars on the pit and on the community which would take years to heal.

The sentences passed at the Nottingham Assizes in June 1937 on the Harworth rioters did not help matters either. There was a widespread feeling in the Federation that Mick Kane and his colleagues had been dealt with unduly harshly having regard to the police provocation which preceded the rioting. In particular, the sentence of nine months hard labour on Mrs Margaret Haymer aroused strong feelings.

The MFGB appealed to the Secretary of State for Home Affairs to reduce the sentences awarded at the Assizes, and eventually some slight remissions were conceded.

Harworth apart, there was very little bitterness in the County. The permanent officials worked together well; and the Council chamber was remarkably free from acrimony.

At its second meeting, held on September 25, 1937, Council received a report from George Spencer on a proposed Pension Scheme for retired members. This was to be financed from the proceeds of the industry and was to provide each retired miner in the County with a pension to supplement his State pension.

The Federated Union had, at the outset, assets valued at £29,824. During the first eight or nine months of its existence its income exceeded its expenditure by rather more than £1,000 a month. Spencer felt that this was just not good enough and he therefore suggested that out-of-work pay should be suspended. At its meeting on July 2, 1938, Council agreed, not without reluctance, that out-of-work pay should be suspended indefinitely. An alternative suggestion that contributions should be doubled, did not commend itself to the membership.

The membership of the Federated Union at its inception is somewhat uncertain: judging by the average weekly contributions, there would appear to have been no more than 16,000, considerably less than the joint membership claimed by the two constituent unions. The Industrial Union alone claimed a membership of 20,000 in March 1937, though this was clearly an inflated estimate used for propaganda purposes. As late as 1944, when the present writer was a branch secretary in the NMFU, some ex-NMA members were refusing to contribute to an organization which had Spencer as its President, and one assumes that this attitude was much more common in the new union's early days.

However, a vigorous membership drive was undertaken and by the end of 1937 some 28,000 people were contributing. There were therefore something like 17,000 persons who were not in membership with the union although some of these (two thousand at the outside) would belong to other unions. The position steadily improved with the passage of time, but even so late as 1943, there were still well over 10,000 non-unionists in the county.

There can be little doubt that the struggle between the two unions, and the manner of its ending had left the rank and file apathetic. Nothing else can account for the high proportion of non-members. After all, to join the Union a man had only to sign a form authorizing the management to deduct the contributions from his wages. Few, if any branches levied the entrance fee specified in the Rules and there were no irksome initiation ceremonies.

Branch meetings were poorly attended, except when matters affecting the members' immediate interests were involved; and at some branches meetings were held only at irregular intervals because of this lack of interest. The Broxtowe Branch for instance, only met twice in 1943.

Again, less than 1 per cent of the membership troubled to pay the political levy of 2s a year. In the nine months ending June 30, 1938, the total income from political levies was £13 12s 6d representing 180-odd members. The Broxtowe Lodge in 1944 had only twelve of its 1,300-odd members paying the levy, and other branches were in much the same position. Prior to the fusion with the Industrial Union, the NMA had had over half its members paying the political levy, and it is difficult to understand this catastrophic decline in political fund membership. Not until the repeal of the 1927 Trades Union and Trades Disputes Act in 1946 was any substantial improvement recorded.

It is only fair to say that all the Permanent Officials, and many of the branch officials, paid the levy; and that the ex-Industrial Union people took part in the propaganda campaign to secure an increase in the political membership. No amount of campaigning was sufficient however, to pierce the barrier of indifference.

A high proportion of non-members, poor attendances at branch meetings and a derisory political levy paying membership are all indicative of an apathetic rank-and-file. In these circumstances, many branch committees became like closed corporations, despatching routine business, drawing their fees, and re-electing themselves to office with monotonous regularity. This then, was the aftermath of dissension. Some of the more restless spirits must surely have longed for the old days of division, when at least it was possible to stir up some interest in the affairs of the Union.

APPENDIX 'A'

Draft heads of an Agreement for the amalgamation of the Notts. and District Miners' Industrial Union and the Notts. Miners' Association, to be entered into after approval by the respective organizations.

1. The agreement existing between the Notts. Industrial Union and the owners until December, 1943, to be accepted.

2. Mr Spencer to be President of the Amalgamated Union. Allocation of offices of the other officials of the two Unions—of whom two (including Mr Cooper and either Mr Hancock or another) shall be officials from the Notts. Industrial Union and two, Mr Coleman and Mr Bayliss, shall be officials from the Notts. Miners' Association—to be the subject of agreement. Mr Ben Smith to be compensated for loss of office and Mr J. G. Hancock, if resigning, also to be compensated. Remuneration and conditions of employment of present full-time officials of both Unions to be not less favourable than at present and their period of employment shall be not less than ten years with the Amalgamated Union except in the case of serious misconduct in respect of their employment.

3. Mr Spencer to represent Nottinghamshire on the Executive of the Mineworkers' Federation of Great Britain for two years; one of the Agents of the present Notts. Miners' Association to be a member of the Executive for the next year following, after which the election of the representative is to be made by ordinary procedure.

4. For two years the appointment of representatives on local Committees, delegates and local officials is to be roughly in proportion to the membership brought into the Amalgamated Union by the two existing Unions. The actual allocation to be settled by local agreement.

5. Men not employed in the industry are not to hold local office in the Amalgamated Union.

6. Political questions unrelated to the mining industry shall not be discussed at ordinary council, branch or executive meetings, but shall be the subject of meetings called specially for the purpose.

7. The Mineworkers' Federation of Great Britain undertake during the currency of the wages agreement terminating in 1943 to refrain from calling on the Nottinghamshire district to take part in any national stoppage or to support any dispute in another district.

8. The accountant at present appointed by the Notts. Industrial Union for the wages ascertainments to be retained by the Amalgamated Union.

9. The Industrial Union's offices to be retained until satisfactory office accommodation for the combined Union has been provided.

APPENDIX 'B'

Provisional Terms for the Amalgamation of the Nottinghamshire Miners' Association and the Nottinghamshire and District Miners' Industrial Union, as agreed on the 26th May, 1937, between the Mineworkers' Federation of Great Britain, the Nottinghamshire Colliery Owners and the Nottinghamshire and District Miners' Industrial Union.

---

1. The two Unions above referred to shall be amalgamated and shall thereafter be known as the Nottinghamshire and District Miners' Federated Union.

2. The amalgamation shall be effective from the 1st September, 1937.

3. The officials of the Nottinghamshire and District Miners' Federated Union shall be:

President	—	Mr G. A. Spencer
General Secretary	—	Mr Val Coleman
Financial Secretary and Treasurer	—	Mr H. W. Cooper
Agents	—	Mr W. Bayliss, and one other who shall be representative of the Nottinghamshire and District Miners' Industrial Union.

The vacant office of Agent above referred to shall be filled by Mr Hancock should he so desire, but should Mr Hancock decline the appointment, then the vacancy shall be filled by the appointment of another person representative of the Nottinghamshire and District Miners' Industrial Union, elected in accordance with the existing rules of that Union, the election to take place not later than the 1st August next.

4. The appointment of the officials above referred to shall be upon a permanent basis, and they shall be subject to removal only for misconduct in the course of their employment. Such officials shall be entitled to equal remuneration and conditions, upon the basis already mutually agreed between the Parties.

Each of such officials will be entitled should he so desire to an individual service agreement embodying *inter alia* the terms above set out, such an agreement to be prepared by Mr John Forster.

5. Mr Spencer shall represent Nottinghamshire on the Executive of the Mineworkers' Federation of Great Britain for a period of one

year. Thereafter, such representatives as Nottinghamshire shall be entitled to send to the said Federation shall be elected under the ordinary procedure as fixed by the rules of the Nottinghamshire and District Miners' Federated Union.

6. For two years from the date of amalgamation the appointment of representatives of local committees, of delegates, and of other local officials, shall be in equal proportions (that is to say, as to one half to be appointed by the Nottinghamshire Miners' Association, and as to the other half by the Nottinghamshire and District Miners' Industrial Union) from amongst men employed in and about the Colliery. Subject nevertheless to certain persons referred to in the course of discussions between the parties being first offered employment. In the case of the persons to whom reference is made above, an undertaking has been given by Mr Spencer that he himself and the officials of the new Union will use their best endeavours to secure employment for such of them as desire it, either at the colliery at which they were last employed, or if this proves impossible, elsewhere in the Nottinghamshire coalfield. The Nottinghamshire owners for their part undertake that they will give favourable consideration to any representations so made. [This refers to the NMA collectors, and not to the Harworth men as Mr Arnot supposes].

It is suggested that the selection of representatives of local committees, of delegates, and of other local officials, should be effected by draw in the branches concerned.

7. It is understood that the rules of the Nottinghamshire and District Miners' Federated Union shall conform to the rules and constitution of the Mineworkers' Federation of Great Britain and shall be in strict accordance also with the provisions of the Trade Disputes and Trade Unions Act, 1927.

8. The Accountant at present appointed to the Nottinghamshire and District Miners' Industrial Union for the wages ascertainments is to be retained by the Nottinghamshire and District Miners' Federated Union.

9. The Nottinghamshire and District Miners' Industrial Union Offices shall be retained until satisfactory office accommodation for the Nottinghamshire and District Miners' Federated Union has been provided.

10. Mr John Forster, the neutral Chairman who presided over the negotiations from which the above terms have resulted, shall be present with officials of the Mineworkers' Federation of Great Britain, the Nottinghamshire Miners' Association and the Nottinghamshire and District Miners' Industrial Union at the taking over of the Funds of the Nottinghamshire Miners' Association and the Nottinghamshire and District Miners' Industrial Union by the new

union, and any question then arising shall be referred to him for determination forthwith and his decision thereon shall be accepted as final.

Further should any difference arise as to the application of any of the terms herein contained, such difference shall be referred to Mr Forster whose decision shall be final.

11. Providing the foregoing terms are ratified then in so far as there may be any matters still requiring discussion in connection with the new wages agreement to be effected with the Nottinghamshire Colliery Owners, Nottinghamshire Miners' Association officials shall be competent, together with officials of the Nottinghamshire and District Miners' Industrial Union, to consider and deal with such matters.

Provided nevertheless that the 31st December, 1943, shall be accepted without further question as the date to which such new agreement shall continue.

APPENDIX 'C'

Notts & District Miners' Industrial Union, 32 Clumber Street, Nottingham.

Balance Sheet—Six Months ending Dec. 31st 1936

Income and Expenditure for Six Months ending December 31st 1936

INCOME

	Total Amt. Collected £ s. d.	Branch Expenditure £ s d	Sent to Head Office £ s d
JULY	1382 4 6	120 10 8	1261 13 10
AUGUST	1461 17 3	106 17 2	1355 0 1
SEPTEMBER	1582 13 9	181 5 10	1401 7 11
OCTOBER	1414 9 0	101 15 1	1312 13 11
NOVEMBER	1416 17 3	100 12 10	1316 4 5
DECEMBER	1652 12 9	208 4 2	1444 8 7
TOTALS	8910 14 6	819 5 9	8091 8 9

	£ s d
Payment of Wages Ascertainment costs	459 14 0
Co-operative Building Society Interests	75 0 0
Halifax Building Society Interests	61 17 6
Woolwich Building Society Interests	15 0 0
Bank Interests	7 1 0
Balance brought forward	4024 17 2
TOTAL	£12734 18 5

JNO. GEO. HANCOCK, Treasurer,
Miners' Office,
32 Clumber Street,
Nottingham.

NOTE: From this statement, it would appear that the Industrial Union had a paying membership of about 7,000; and not of 15-25 thousand as claimed by supporters of Spencer. The assets were also much lower than was popularly supposed.

EXPENDITURE

	£ s d
To Agent's Salary and Expenses	350 19 6
,, Secretary's Salary and Expenses	251 10 5
,, Treasurer's Salary and Expenses	229 19 3
,, Solicitors	612 7 9
,, Lodge Expenses	529 1 7
,, Wages Ascertainment costs	333 10 7
,, Motor Expenses	198 3 7
,, Council Meeting expenses	153 4 0
,, Doctors	110 14 9
,, Organizing expenses	99 5 9
,, District expenses	89 8 8
,, Conference expenses	85 17 0
,, Rents	55 6 9
,, Printing expenses	52 6 7
,, Executive Committee expenses	49 16 7
,, Inspection expenses	46 8 6
,, General Rate	32 5 4
,, Telephones	29 17 7
,, Auditing	15 15 0
,, Bank Charges	6 9 6
,, Water expenses	1 4 0
,, Electricity	1 12 5
TOTAL	£3333 16 11
,, Out of Work Pay	2746 11 0
,, Pension Fund	613 0 0
,, Wages Board	45 0 0
,, Balance carried forward	5996 9 11
TOTAL	£12734 18 5

APPENDIX 'D'

Notts. Miners' Federated Union
Pension Scheme

Miners' Offices,
Nottingham Road,
Old Basford.
January 10th, 1938.

Dear Sir,

The Executive Council has recommended that a Council meeting be held to consider what steps, if any, should be taken to establish a pension scheme.

Some years ago the old Association consulted an eminent actuary and fortunately we have information which enables us with some accuracy to formulate such a scheme.

The information referred to shows:

1. That if the average age of men working in the pits of Nottinghamshire was 35 years a sum of £880,000 would have to be provided, and 6¼d. a week paid per member in addition, to secure a pension of 5/- a week.
2. If the average age was 45, the lump sum required would be £1,340,000 and 1/0½d. a week per member would have to be paid. This is for a pension of 5/- a week.

It is quite obvious that we cannot find such large sums as mentioned to make the scheme financially sound, unless a scheme is adopted which takes advantage of rising wages to increase the weekly contributions.

The first definite point we wish to make is:

That if a pension of 10/- a week is to be paid to all contributors a very large sum of money has got to be found.

It is our belief that this sum can be accumulated over a number of years provided two things:

1. That there is a continuation of the standard of wages approximating to that of the present time.
2. That the workmen are prepared to make a contribution out of their wages on a percentage basis.

As we have no desire to deceive anyone, we wish to make it perfectly plain to accomplish a scheme of this character the young men, and those who are getting the highest wages shall, having regard to all the circumstances, make the highest contributions.

The scheme we recommend is one in which the workmen would pay 1% upon basis wages when wages were at the minimum; that is to say instead of receiving basis wage plus 38% they would receive basis wage plus 37%.

As wages increase we recommend that the percentage payable for pensions increase in the following ratio:

Percentage payable on basis	Percentage to Pension Scheme
38%	1%
50%	2%
60%	3%
70%	4%
80%	5%
90%	6%

Plus all broken percentages

and 1% for every further 10% on wages basis

If this scheme was adopted and there was an average percentage on the present scale through the year of 4% it would give us £144,000 per annum.

If this scheme was at once put into force we could make provision to pay pensions immediately to all men attaining the age of 65 at the present scale, and at a later date when funds were available we could reconsider the amounts to be paid.

We would like to point out as we have stated above, that the information we have in our possession, if 10/- a week had to be paid in pensions, the following amounts would have to be paid as contributions in addition to the large capital amounts required.

Member's present age	Weekly premium to age 65
25	7½d
35	1/0½
45	2/1
55	5/8
65	a sum of no less than £213 12s would have to be put on one side

It is quite obvious in view of this information that a pension scheme can only be established if the young men are prepared to assist the elder men, and those getting large wages are willing to make some contribution to those who are getting a less standard.

If the scheme was in force and a man had £5 basis wage, when percentage was 80% he would pay 5/- a week, but he would take home £8 15s., whereas a man whose basis wage was £2 would pay 2/-, but he would take home only £3 10s.; so while a man with higher wage would pay more, he would have a far greater sum to pay it with, and if he is a man of 55 years of age and the man receiving £2 is a man of 25, the man who is paying the 2/- is actually paying more than the man paying 5/-, having regard to the actuarial principles.

We are looking at this point from a humanitarian point of view as well as an actuarial, and we say without equivocation unless the young men and the higher paid men are willing to make some contribution for those who are rather less fortunate or advancing in years then this scheme cannot be carried out; but we believe there is sufficient humanitarian feeling in the county amonst the younger men to justify our putting this scheme before you and recommending you to accept it, believing that it will form the basis of a scheme that will provide the younger generation with a pension when they attain pensionable age.

We shall make provision in the Rules which will be formulated for contributors to continue their membership if incapacitated through sickness or accident, also for contributors continuing their contributions if they leave the industry.

G. A. Spencer, President.
Val. Coleman, Secretary.

APPENDIX 'E'

This letter from G. A. Spencer to the Nottinghamshire Coalowners gives some idea of the pressure he put upon them to concede advances to his members. It will be seen that he used the NMA as a 'Bogey' to force the issue.

MIDLAND COUNTIES COLLIERY OWNERS' ASSOCIATION

Wilderslowe,
Derby.

PRIVATE & CONFIDENTIAL

27th February, 1936.

Dear Sir,
Wages Position

By desire of Mr G. A. Spencer I forward herewith for the private and confidential consideration of each of the Nottinghamshire Coal Owners, copy of a letter received from him bearing date February 26th, 1936, which is given below.

The Chairman particularly desires that each Nottinghamshire Coal Owner should be represented at a Meeting to be held on Monday next, March 2nd, 1936, at 11.15 a.m., at the Victoria Station Hotel, Nottingham, and that the representatives should have full power to vote.

The subject is one of importance, and a decision is requisite for the Meeting of the Notts. Wages Board to follow on Monday next.

Yours faithfully,
William Saunders.

To Coal Owners of Nottinghamshire.

COPY   32, Clumber Street,
                                                    Nottingham.
                                             26th February, 1936.

William Saunders, Esq.,
Wilderslowe,
Derby.

Dear Sir,

I shall be obliged if you will circulate this letter to each of the
Colliery Companies within the district of Nottinghamshire.

Since our Meeting of the Wages Board on Monday, when the
financial position of the Industry as relating to Nottinghamshire
was considered, I have been able to obtain further information from
our auditors and that information has led me to believe that the
decision there arrived at, at the said Meeting, was not based on the
most up-to-date information which has since been available. I am
now informed that the profits for January will be approximately
£230,000 on 1,454,000 tons, which represents a profit of 3/2d per ton.

For the purpose of comparison I should like to point out to the
Owners that the profit for the month of January 1935 was £107,000
so that the increase in profit for January 1936 is, on these figures,
£123,000 and that after the 1/-d and the 6d have been paid to the
workmen. To put it in another way, it means that the increase in
the profits for January 1936, compared with January 1935, is £161,000
out of which £38,000 has been paid to the workmen in increased
wages, and, even upon this £38,000, the Owners have received—
rightly—their profit of 15%. In view of this large increase in profits
I should be much obliged if the Owners would again meet before
next Monday for the purpose of re-considering their position not to
permit the Workmen to participate in these profits for the month of
March. The answer has been given that, to do so, would be a violation
of the terms of the Agreement and that the extra profits which have
accrued are due, not to the increase in the extra money Contractors
have paid, but due to the increased price which has been paid for
coal which was not contracted for. I am quite prepared to admit such
a position but I should like to point out that I should neither be able
to convince the Workmen nor our opponents—The Nottinghamshire
Miners' Association—nor the general public on the distinction which
you make between profits accruing from contracted coal and non-
contracted coal; that is a distinction that neither of them will
appreciate. The one thing that they will consider is that there has
been an enormous increase in the profits and that under the circum-

stances, having regard to the promises which have been made by the Owners to the Workmen and to the Public, they should be entitled to participate. If they are not permitted to do so I am afraid that no amount of preaching of mine, or anyone else's, will convince them that there is any spirit of co-operation between us and to avoid any misunderstanding, or to avoid any conflict of opinion or of interest I beg the Owners to meet again for the purpose of giving the Workmen's side an opportunity of re-stating their case. If this is denied I must, speaking personally, seriously consider the question of resigning my position.

I should here like to state what is my personal view of the situation, which I know is endorsed by our side of the Board and by the members of our organization. I feel that if this happy financial position which has arisen in Nottinghamshire was the position of the whole of the coalfield of Great Britain that the request which has been made on behalf of the Workmen would be immediately conceded, for be it remembered, and I again reiterate that we do not want a single penny that belongs to the Owners under the terms of the Agreement, but we do ask them to share the profits with their Workmen now to the extent of, say, 5% advance, which they can readily do, knowing that recoupment is now certain so far as deficiencies are concerned. I say it would be granted if this was the general financial position of the country because of the power which the Federation alone possesses and if what I am surmising is true I should like you to see that it must naturally follow that the Employers are taking advantage of our isolated position: Moreover I should like them to understand that we have sufficient difficulty and worry in facing the hostility of the other side, and, therefore, it has been regrettable to me that the Owners could not, at our last Meeting, go out of their way a little to concede a point which is not detrimental to their interests. I have had endless worry since our Meeting, and I regret very much that it has been necessary to write this letter, but I say, most respectfully, that if there is not such a Meeting as I suggest I must seriously consider my position, and my conviction, for the moment, is that it would be better for me to resign.

Yours faithfully,

Geo. A. Spencer.

APPENDIX 'F'

THE HARWORTH SENTENCES

The trial of the Harworth rioters opened at the Nottingham Assizes on Wednesday, June 23rd, 1937, and continued on the two

13. Re-union 1937
Left to right: G. A. Spencer, Val Coleman, H. Cooper, B. Leese (Clerk),
W. Bayliss and H. W. Booth, outside the Miners' Offices, Basford

following days. The prisoners, who wore labels for purposes of identification, were:

George Barker (aged 28) William Carney (35)
George A. Chandler (44) Frederick Halliwell (44)
Frank Jobson (39) Michael Kane (39)
Thomas Morris (36) Bernard Murray (25)
Luke Parkinson (30) William Pank (40)
Thomas Richardson (28) John H. Smith (33)
Thomas Smith (43) John A. Wilson (38)
Albert E. Ridsdale (38) Henry Billam (44)
Mrs Margaret Haymer (31)

The main charge against them was one of riotous assembly with other persons unknown.

It was stated in evidence that a crowd of 1500–2000 people assembled on April 23rd at the time when the 'chain gang' was being marched to work under police escort. Stones and bottles were thrown and buses carrying strike-breakers to work were attacked and their windows smashed.

Garden walls were said to have been knocked down to provide ammunition, and bricks were thrown through the windows of houses occupied by men at work.

Ridsdale was alleged to have led an attack on the Miners' Institute; whilst Mrs Haymer was said to have struck a man named James Taylor with a bottle.

The accused denied having seen sticks or bottles used and they maintained that prosecution witnesses were guilty of wild exaggeration. A number of witnesses testified that Mrs Haymer did not have a bottle in her hand at any time during the disturbance; and indeed one witness admitted to having struck James Taylor, but his evidence was discounted.

Michael Kane, who was described as the ringleader, was sentenced to two years hard labour; there were two further sentences of 15 months, one of 12 months, six (including Mrs Haymer) of 9 months; one of 6 months and one of 4 months. The other accused were bound over to keep the peace.[9]

T

CHAPTER 16

'ONE BIG UNION'

The Union's proud motto 'United We Stand, for Unity is Strength' once more had meaning. It was however, a parochial unity. George Spencer had no sympathy with those who wished to create one big national union of mineworkers in place of the loose federation of autonomous district unions; and for the moment, his views went unquestioned in the new Union.

However, in the Mineworkers' Federation at large the tide was running strongly in favour of national organization with its corollary: national negotiating machinery. It will be remembered that this issue was originally raised during the First World War when, as a matter of convenience, the Miners' Federation EC had acted as the principal negotiating body with the Mining Association and the Government. South Wales had campaigned for the continuation of national negotiating machinery in peacetime, and this had become the official policy of the MFGB. At that time, George Spencer had been the most vociferous opponent of this policy and it was his willingness to countenance District negotiations that took him out of the Federation in 1926.

The owners for their part, had refused to accede to the Federation's demand for national negotiations. They believed that the wage-rates paid in a district should be governed by that district's ability to pay and not by some objective assessment of the worth of particular kinds of labour. Further, they believed that national negotiations invited political interference and that this interference could open the door to governmental control. As we saw in our last chapter, the MFGB had, by 1935, sufficiently recovered the strength it lost in 1926 to secure the establishment of a 'Joint Standing Consultative Committee' with power to discuss wages questions in a general way. To the militants of the Mineworkers' Federation, this was merel the opening shot in the campaign. Next, they intensified their attempts to bring about the creation of a National Union of Mine workers.

Prior to its fusion with the Industrial Union, the NMA had supported the militants on this, and every other important issue. Now however, its militant interlude was over. The Federated Union was in future to be numbered among the moderates as the old NMA had been in Hancock's heyday.

In the Minutes of the NMFU Council Meeting held on Saturday, June 4, 1938, we find this entry:

'The President made reference to the Annual Conference, also to the Federation EC Report, which contains a recommendation for the formation of one big Miners' Union.

'It was resolved the Conference Agenda be referred to the delegation, with instructions to vote against one big union.'

However, at the 1938 Annual Conference no firm decision was taken and it was not until the 1939 Conference, which opened at Swansea on Monday, July 3rd, that George Spencer made Nottingham's attitude clear.

He started by saying that the MFGB was an 'efficient medium of expression' and that it was not the looseness of the organization which brought defeat in 1921 and 1926 but rather the 'overpowering economic difficulties of the moment'. He went on:

'If history has not done for us anything else, it has taught us this, that our power is limited by the economic possibilities, not only of the industries themselves, but of the passing years and of the phases of those years. It has been said in this Conference today, or during the week, that some of the advances that the Districts have got have been by the action the Federation took when it started its campaign. I am certain that thoughtful men of this Conference would never believe that idea for a moment. It was because there was the beginning at that time of the armament policy, and a consequent improvement in the economic capacity of our industries to pay. I would only wish from this rostrum that the lot of Nottingham had been the lot of the whole Federation. Our average wage in 1936 was about 10s 7d. Our last returns . . . [show an average of] 15s 8d a shift. Do you think that is because of our organization and its power, or do you think it has been owing to my eloquence, or that it is our unity, that has obtained that? Not at all . . . I am here not to be condemning the Owners all along . . . my life is not long, but what years belong to me will be spent in working with the Owners to get the best out of industry, and when we have got it, with the united effort and co-operation of the workmen, I demand our fair share. That is my policy. Other people can have theirs; I shall not criticize them . . . in emphasizing the point I was trying to make, the 15s 8d is not through the strength of my organization. True, without the organization we could not get it. I know that. There is not sufficient benevolence in the Owners to do it without some organization. But the organization—and that is the point I am trying to make—however powerful it may be, whether it is the Trade Union Congress, or whether it is this Federation, or whoever it may be, it will not succeed beyond the capacity of the district or the industry to pay.

'. . . Let us suppose there is a fall in the market; there is a psychology in the Federation for a strike; we shall involve the Districts in a turmoil, and instead of making for unity at that moment we shall have all the elements for breaking us up again.'

He went on to say that he would resent any national officer of the Federation who was unaware of local conditions, attempting to negotiate for his district, and he ended:

'. . . And there lies our danger—encouraging the smaller Districts to think that they can secure by financial assistance [for strikes] what the limitations of the capacity of their own industry and their own district will not give. . . . We will give to the Federation anything it likes in money to make it efficient as a means of giving adequate expression to national questions, and for providing all the sinews for getting information; but for building up emergency funds to make grants we say "No, let each District provide", for it is the limitation of its funds sometimes which is the measure of its wisdom, and consequently if you break down that wisdom by in anyway removing the necessity for it to be discreet in its actions, you would be taking the first step, in my opinion, towards doing a great harm to the District, and finally it would bring the elements of disintegration into the Conference again.'

Had the country remained at peace, the development of national negotiating machinery would probably have been a slow and painful process having regard to the opposition of the owners from outside the Federation and of Spencer from within. However, the Second World War, like the earlier conflict of 1914–18, acted as a catalyst.

Shortly after the outbreak of the war, the Secretary for Mines wrote to the MFGB and the Mining Association to suggest that the two sides should discuss 'the question of increased production of coal to meet the requirements necessitated by the war'.[1]

At a meeting of the Joint Standing Consultative Committee held on September 21, 1939, the Federation delegates readily pledged the co-operation of their members in the drive for increased production. At the same time they asked that for the duration of the War, the Mining Association should once more be empowered to negotiate on wages issues. In particular, they suggested that increases in wages to meet increases in the cost-of-living should be negotiated nationally. Within a week, the district colliery owners' associations had agreed to allow the Mining Association to negotiate for them. Shortly after, a flat-rate increase of 8d a day for men and 4d a day for boys was agreed on.

At a Special Conference held on October 27, 1939, George Spencer expressed his misgivings at the resumption of negotiating power by the Executive Committee of the Federation. He said:

'Every district has a right to give that power and authority to the Executive if it likes, but my submission, without labouring the point, is that the Executive have no right to assume that authority.'

He felt that the Executive should confine itself to advising districts on the size of the advances they should seek having regard to movements of the cost-of-living index; and to conferring with the owners and the Government on increases in the selling price of coal.

He went on to attack the principle of flat-rate increases. Each district, he thought, should be free to say whether it preferred flat-rate or percentage increases. He rejected the argument that flat-rate increases were better because the lower-paid men had to go to the same market as the higher paid. If that was to be the criterion then it followed that all workers should be paid the same wage irrespective of skill since they always had to go to the same market.

Spencer went on to point out that while the average increase in wage-rates in the Nottinghamshire coalfield during the June quarter of 1939 over the 1914 wage was 133 per cent; for the lowest paid grade of labour the increase was 195 per cent at soft coal pits and 180 per cent at top hard pits. The differentials had thus been narrowed considerably, and the wages of the lower-paid man had been so lifted that '. . . the higher paid man has had to make a contribution to his wages'. Spencer envisaged a situation developing where the differentials had been so narrowed that the coal-face men would say, 'Let the other man come and get the coal; the differentiation in wages is so narrow now that he might as well come and have the risk with me'.

Replying to Spencer, Arthur Horner of South Wales argued that the lower paid men needed the cost-of-living increase more than the higher-paid men. He went on to deal with the deeper issue of national organization. There was he said, a 'fundamental difference between some of us and Mr Spencer . . . I want national organization. I believe in national control of the wage policy. I believe in using every possible situation to unify the conditions of the miners of this country. I do not think it is right that the accident of geography should determine that Welsh miners should get two-thirds of what the miners in the Midlands are getting. I do not even believe that the exigencies of war should produce a position where export districts such as I come from might entitle us for a period to wages far in excess even of the inland districts. Our national policy has been to try to secure the maximum national control. Someone may say "Yes, you have got it, but it is only for the period of the war." I put it to you, that if it is good to get it for ever, it must be good to get it for three years or whatever the war period may be.'

The debate continued at subsequent conferences. At the 1940 Annual Conference which opened at Blackpool on July 15th the exporting districts expressed concern at the loss of their continental markets. They were afraid that their pits would be on short time whilst those of the inland districts were overburdened with orders, and they demanded that the trade should be equally shared. To use the words of Mr J. E. Swan of Durham:

'The policy we are putting forward is an economic policy, and as well as being economic it is a moral policy, that whatever amount of coal production is required, it should be shared. No man in the Midlands or elsewhere can justify a system which allows some miners to be over-worked whilst others are under-worked.

'. . . As has been said, we are all in to win the war, and neither Yorkshire, Nottinghamshire and Derby nor the Midlands can win the war alone. Durham, South Wales and Scotland are of equal importance. In unity we can proceed to victory, and in unity we can share whatever trade there is for the benefit of all.'

George Spencer would have none of this. He alleged that the northern districts, by reducing their prices, had taken trade amounting to nearly 2,000,000 tons of coal a year away from the Midlands. He went on:

'Remember this, for years in the Midlands we have averaged less than 4 days per week—3 and 3½—while some districts who had stolen our coal made 6. There was no question raised then about equity of tonnage to the districts; none whatever. . . . Remember always this, that the competition of coastwise trade hit us in the Midlands, especially in our house coal trade, to a degree that is not understood by people who have not given thought to it, and by districts that have depressed their wages 33 to 40 per cent below the standard of wages that exist in our district. I think it is a fair inference to draw that some of the districts which are now complaining took a very great deal of our trade by underselling us and by depressing their own wages.'

Later in his speech Spencer produced a document prepared by Lord Hyndley (Commercial Adviser to the Mines Department) which he suggested was Government inspired. This contained the following proposals:

'Lancashire and Cheshire and the Midlands Amalgamated District coal should not be shipped to Ireland or Northern Ireland. . . .

'Moreover, coal should be moved to London and the South. The Southern Railway, the Great Western Railway, should draw their supplies of locomotive coal from South Wales. . . .

'An unspecified number of public utility and industrial consumers now supplied by the Midlands Amalgamated District should be supplied by other Districts. South Wales coal should go to Lancashire. . . .

'Northumberland and Durham coal should be moved in large quantities, in large tonnages, northward.

'Northumberland and Durham coal should move in large quantities into North Yorkshire replacing coal normally supplied in that part of the country.'

This document was something of a mystery. Spencer stated that: 'This document came into my hands I scarcely know how'; although the chances are that he knew perfectly well. The picture of a white-haired George Spencer stumbling innocently across an alarming document is touching but unconvincing. In all probability he obtained it from one of his coalowner friends. As Arthur Horner said later in the debate: 'This resolution from South Wales has not come out of any mysterious coalowner's bag. The only document I know of which is likely to have come out of a coalowner's bag is the one quoted by Mr Spencer here this morning.'

Spencer concluded by saying that his district would not be opposed to some temporary arrangement to keep the pits of the exporting districts at work, but he wanted firm assurances about the future. He also insisted that the arrangements should be made between districts, and not imposed by the Federation.

It was finally agreed that this matter should be remitted to the Executive 'in order that it shall be considered in the interests of the Federation as a whole'.

In the event, all coalfields found ample outlets for their product in the home market; and indeed, before long British miners were being exhorted to produce more and yet more coal to keep the wheels of industry revolving. When it became apparent that exhortation alone was not enough, the Government sought to introduce a measure of compulsion. In the early part of 1941, an Essential Work Order (S.R. & O. 1941, No. 707) was drafted. The object of this Order, according to Ebby Edwards, was to: '. . . prevent the loss of produc-

tion owing to the transfer of labour to other industries, or by unnecessary absenteeism and other behaviour which impedes effective
production.'

The MFGB Executive tried to obtain a comprehensive National Wage
Agreement and national control of the mines as a *quid pro quo* for
supporting the measure of industrial conscription contained in the
Order, but the owners and the Government were opposed to this.
As Ebby Edwards reported to a Special Conference held on May 8,
1941:

'I would be misleading the Conference if I said there was the
slightest hope of the Executive going to negotiate . . . anticipating
that any further negotiation would get included in this Essential
Work Order either a national board or a national increase of
wages.'

This displeased a great many delegates, who felt with Mr J.
Pearson of Scotland that:

'This Conference should simply turn down the Essential Work
Order, and we should go forward for the three points which are our
basic demands, namely, the creation of a National Board, a
satisfactory guaranteed weekly wage, and 100 per cent organization
in every way.'

On the other hand, George Spencer opposed the creation of a
National Board, and he attacked the guaranteed wage provisions in
the Essential Work Order. The Order provided for a levy to finance
guaranteed wage payments made to men able and willing to work,
but who were prevented from working full-time by circumstances
outside their control. This evoked from Spencer the following
declaration:

'Now I come to deal with the question which has called forth
our opposition, and I am not a man given to exaggerated and extreme language, but I will say this in this Conference, that if the
wages of the workmen, honestly earned, in our district, have got to
be taken from them in the form of a pool to be distributed in some
other district, then in face of the Government, in face of the three
months' imprisonment or the £100 fine, I would advise our men not
to accept it. I do not say that boastfully. I think that is an im-
position, and something which no district should be asked to
accept. I say that in no other industry where you have this guaranteed week is one body of workmen being asked to make any
contribution to some other body of workmen.'

Spencer suggested that the guaranteed wage payments should be financed by an increase in the selling price of coal, and should not be dealt with through wages ascertainments, since in the latter case, 85 per cent of the money would, in effect, be found by the workmen.

Mr J. A. Hall of Yorkshire felt that they should demand improved wages and conditions as compensation for the liberties they were about to lose. He also pointed out that, in order to guarantee owners' profits, the Government had permitted an increase in the selling price of coal, and he thought that the guaranteed wage should be similarly financed.

On the morning of May 9th, the second day of the Special Conference, the following resolution was carried by 464 votes to 100, South Wales alone opposing it:

> 'That the Conference stands adjourned for a meeting of the Executive Committee with the Government, to impress upon the Government the necessity of their consent to our request on the three points—the National Board, increased wages, and non-unionism—if needs be in or outside the Order.'

The Executive met representatives of the Government at 11.30 am and reported back to Conference in the afternoon of the same day. The President of the Board of Trade had handed to them a typescript containing the following points:

> '(1) The Government regard it as a matter of vital national importance that the miners' representatives should today recommend the acceptance of the Essential Work Order.
> '(2) The Government regard the subject of rates of wages as a matter for negotiation under the existing machinery and could therefore not agree to introduce legislation which would compel the establishment of a National Board.
> '(3) The Government understands that if so requested the owners are prepared to enter at once into these negotiations, and the Government undertakes to do all possible to ensure that they are carried through speedily.'

After a lively discussion, Conference adopted the following Resolution by 370 votes to 194, Nottinghamshire voting with the majority:

> 'That the Conference strongly protests against the owners and the Government failing to appreciate the need for increased wages in the operation of the Essential Work Order, but in view of the serious war situation the Conference recommends the Consultative Committee be given power to examine, seek to amend and apply the Order in terms applicable to the mining industry.
> 'Further, the Conference recommends the Executive Committee

to press for fundamental changes in the industry having regard
to the wages position.'

Subsequently, on Wednesday, May 21, 1941, the National Joint
Standing Consultative Committee met '. . . to consider a request
made by the Mineworkers' Federation of Great Britain that every
adult mineworker should have a weekly minimum wage of not less
than £4 (Four Pounds) per week'. It is clear that the Executive Committee were attempting to obtain their 'National Board' by the back
door, as it were. To concede a national weekly minimum wage the
Mining Association would have had to make considerable inroads
into the autonomy of its constituent district associations. This the
Association was not prepared to consider. Its President, Sir Evan
Williams pointed out that the Agreement of March 1940 provided
that:

'District wage arrangements shall continue to operate during
the War subject to mutually agreed alternatives, but increase of
wages necessary to take account of the special conditions arising
out of the War, and particularly the increased cost of living shall
be dealt with on a national basis by means of uniform flat-rate
additions.'

Sir Evan characterized the proposal for a national minimum wage
as '. . . entirely novel and for which there did not seem to be any
occasion at all'.

The owners were, however, prepared to consider a flat-rate payment, in addition to the cost-of-living bonus, for those men who
made full-time. This was intended as an inducement to men to refrain
from absenting themselves from work without good reason. The
original offer was one of 6d a shift conditional upon the Government's
agreeing to a corresponding increase in the selling price of coal.

The MFGB Executive then reluctantly agreed to drop its demand for
a national minimum wage (some of the districts, including presumably, Nottinghamshire, were opposed to the idea anyway) and to
press instead for an attendance bonus of 1s a shift. This was eventually
agreed upon and a corresponding rise in the price of coal—10d per
ton—was authorized by the Government to operate from June 1,
1941.

However, during the negotiations on the conditions which were to
govern the payment of the attendance bonus, the weakness of the
Federation representatives' position manifested itself. As Ebby
Edwards said at a meeting of the National Joint Standing Consultative Committee held on June 5, 1941:

'We are both in a very, very unfortunate position because of our own machinery and constitution under which we have to act. With the local district autonomy— and the same applies to your constitution—we are both trying to act nationally with autonomous district decisions against national procedure.'

Then followed this exchange:

'SIR EVAN WILLIAMS: Well we are in the position that whatever we agree to with you will be accepted by the industry as a whole. We are not going to be put in the position of making conditions here with you and you going back to your people and being turned down.

'MR EBBY EDWARDS: But we were turned down before we went to our Executive Committee.

'MR W. LAWTHER: These things we have asked for are the result of the discussions of the Executive Committee.

'SIR EVAN WILLIAMS: Am I to understand that these are the conditions put forward from the whole of your Executive Committee, and as far as you are concerned, have you any power to negotiate and settle the position?

'MR W. LAWTHER: No. We have to report back.

'SIR EVAN WILLIAMS: Well that puts us in a very unsatisfactory position. We ourselves are in a position here to decide what we think best and it will be agreed to by our people.'

The Executive Committee, at its meeting on June 6, 1941, did however, take responsibility for agreeing to the conditions governing payment of the attendance bonus which was to operate from June 1st. When the Annual Conference met on July 14th, this decision came under heavy fire from a number of delegates and the Executive was instructed to '. . . approach the Owners with a view to the withdrawal of the conditions attached to the attendance bonus'. In particular, delegates were opposed to the condition that, in order to qualify for bonus, a man must be 'capable of and available for work throughout his normal working hours during the week'.

At this conference the Executive sought to strengthen its hand by presenting a motion advocating the setting up of a National Board composed of an equal number of representatives from the Mining Association of Great Britain and the Mineworkers' Federation of Great Britain with power to decide 'all matters which directly or indirectly affect the wages, conditions, and safety of mineworkers'. This motion contained the revolutionary proposal that:

'Present district ascertainments for wage purposes shall be supplemented by a National Ascertainment for the purpose of making an equal distribution throughout every district of all wages surpluses over the new standard wages. . . .'

This motion was strongly attacked by George Spencer who said:

'This idea of control which has been mentioned is purely a red herring drawn across the track of Conference at this time for the purpose of tapping the revenue of the best districts and transferring it to the others.'

He pointed out that the average output per man shift at the face in Nottinghamshire was 84 cwt. against a national average of 59 cwt. and he asked delegates whether they thought that Nottingham miners '. . . look on and the coal puts itself into the tram or on to the conveyor'.

He went on to remind Conference that a number of districts had, earlier in the day, castigated the EC for misusing the power delegated to them in the matter of the attendance bonus and he argued:

'Is not that strong evidence that it is a most dangerous course to vest the Federation with the power to negotiate on behalf of all districts irrespective of the economic capacity of the districts to produce coal?'

He again insisted that Nottinghamshire's 'economic capacity' was not determined by geographical or physical features and he submitted the following evidence:

'We are the third lowest in price in Great Britain. There are districts getting 2s 6d and 3s a ton more than we do. We are getting seams just over 2 ft thick as well as you.'

He could understand leaders in other districts trying to raise the level of wages of their members and he had no objection to that—provided that their improvements were not gained at Nottinghamshire's expense. He concluded:

'I am telling you plainly that unanimously in my Council last Saturday this resolution was turned down, and I will be no party to taking from the earnings of the workmen in Nottingham a single penny piece to pay not only 90 per cent to the workmen, but in some mysterious way to pay 10 per cent to the owners.

'This is absolutely a new departure. It is a new principle in the Federation—a principle in which we are asking the workmen of one district to help to pay the profits of another district. If that is not the right construction upon the 90 per cent and the 10 per cent tell me what it is.[2] And so far as we are concerned, I have no hesitation in saying that we offer the strongest opposition to these proposals, that we will not accept them, and that if we were therefore driven from the Federation, we should have to go. If they are put into force, let no one have any illusions about it, and if we have to be driven from the Federation, we should be driven, because we shall not accept proposals of this character.'

Although Spencer was the only delegate to make a root-and-branch attack on the motion, several others criticized it in detail, and it was agreed that the whole of the motion should be remitted to the Executive Committee for further consideration.

The advocates of national control found an ally in the increasing demand for coal at a time when output was falling. So serious was the position that, in 1942 the Government decided to assume control of the production and allocation of coal. A White paper (*Commd.* 6364) setting out the Government's proposals was presented to Parliament by the President of the Board of Trade on June 3, 1942. This provided for the setting up of a National Coal Board with power to plan production, and consider various matters relating to manpower, supplies of materials, welfare facilities, and so on. In addition, Regional Controllers were to be appointed:

'. . . to whom will be delegated the Minister's powers to control colliery undertakings and give directions to the management, these Regional Controllers, advised and assisted by Regional Coal Boards, to have full responsibility for the conduct of mining operations in their Regions.'

The Government felt however, that:

'. . . the success of the proposed National Coal Board as a body for increased production would be gravely prejudiced if it were associated in any way with wages questions.'

At the same time the MFGB were pressing their claim for a guaranteed national minimum wage, a claim which the Mining Association were not prepared to concede. The Minister of Labour and the President of the Board of Trade (Ernest Bevin and Hugh Dalton) therefore appointed a Board of Investigation under the chairmanship of The Right Honourable Lord Greene, OBE to consider the immediate wages issue and to 'submit recommendations for the establishment of a procedure and permanent machinery for dealing with questions of wages and conditions of employment in the industry'.

The Board of Investigation in its report dated June 18, 1942 came down heavily in favour of the MFGB. They recommended an immediate flat-rate increase for adults of 2s 6d a shift, a guaranteed national minimum wage of 83s a week for underground workers and 78s a week for surface workers, and an output bonus to be calculated on a sliding scale. These recommendations were subsequently adopted by the Government.

The immediate wages issue disposed of, the Board of Investigation turned its attention to the establishment of conciliation machinery. Discussions on this subject between the Mining Association and the MFGB were already being held when the Board of Investigation presented the two sides with a draft scheme of its own. This draft was used as the basis of further discussions between the Joint Standing Consultative Committee and the Board of Investigation and eventually an agreed scheme emerged.

This provided for the setting up of a National Wages Board to consider all questions of a national character affecting the wages and conditions of employment of mineworkers. This National Board was to be composed of two bodies: a Joint National Negotiating Committee with equal numbers of owners' and miners' representatives and a National Reference Tribunal consisting of three independent persons to be nominated by the Master of the Rolls. Any question which could not be resolved by the JNNC would be referred to the National Reference Tribunal whose decision would be final.

Similarly, each district was to have a District Conciliation Board representative of the two sides to settle questions of a local character. Any district questions which could not be resolved by the Joint Board were to be referred to an independent District Referee whose decisions on such questions would be final.

A special conference of the Federation held on Friday, January 22, 1943 at the Conway Hall, London recommended acceptance of the Scheme. At this conference, Arthur Horner, speaking for the Executive, said that the Federation:

'. . . having compelled the Government of the day to realize the necessity for national organization for the conduct of production and wage relations, will at long last appreciate that this has certain implications for us, and that this organization might have to change its form so as to become more compact, more coherent, more united, and so able to take its full part in the conduct of the industry and in the affairs governing the relations of workmen and employers.'

At the meeting of the Joint Standing Consultative Committee held on February 25, 1943, it was reported that the district miners' and owners' associations had unanimously agreed to the setting up of the conciliation machinery. The Federation's representatives on the Joint National Negotiating Committee, George Spencer among them, were appointed on Thursday, April 8th and the Committee held its first meeting on July 8th.

Although Nottinghamshire voted in favour of establishing the national conciliation machinery, the next step proposed by Arthur Horner, which amounted to the formation of one National Union in place of a loose Federation of districts, was by no means universally approved. The rank-and-file were mainly in favour, it is true, but Spencer and his friends were opposed to the whole idea. They felt that the formation of a National Union of Mineworkers would lead to the lowering of the wage standards of the more prosperous counties, and they illustrated their argument by referring to the way in which the Coal Charges Fund worked. This Fund was fed by a levy on each ton of coal produced, and it was used, in effect, to subsidize the less profitable districts. Referring to a proposed increase in the national minimum wage at a Conference held on Thursday, January 27, 1944, Spencer had this to say:

'. . . I have no objection to all sections getting a reasonable wage. But speaking as a leader of Nottinghamshire, I have a pronounced objection to either one district or another district getting a substantial increase in their wages—not if that is denied to me, but further, and more important—when they have got to call upon my district to find the money to pay those wages. That is the ground of my objection, and that is the gravamen of my charge against those who have formulated that policy.

'The question I would like to ask is this: there is to be a further 3s per ton put upon coal. So far as I have been informed—probably wrongly informed—that 3s per ton is going to be diverted from the Ascertainments into the Coal Charges Fund. Let us assume for the moment that I am correct. Nottinghamshire turns out approximately 30 cwt. per man per day; 3s per ton is 4s 6d per day. 85 per cent of that giving the owners their 15 per cent share, will mean that in Nottinghamshire the man who goes six days will be contributing 23s 10d per week, and if he is fortunate enough he will probably get the two half-crowns for Friday and Saturday as an extra attendance bonus.'

By this method of calculation, the Nottingham miner would therefore be contributing a net figure of 18s 10d a week extra in supplementation of the wages of the lower paid men in other coalfields. Spencer elaborated this argument in speeches and newspaper interviews, but it appears to have had little effect among the rank-and-file members of his district.

The supporters of the proposal to convert the Federation into a National Union of Mineworkers were active in Nottinghamshire as they were in other counties. In the winter of 1943–44 a campaign was organized by the Communist Party to counter Spencer's arguments. Most of the local leaders who had been associated with the NMA were equally strongly in favour of the formation of one big union whilst those who opposed it were, in the main, former Spencerites.

In its report to the 1943 Annual Conference, the MFGB Executive Committee had presented a draft constitution for a National Union of Mineworkers. This provided that the industrial work of the proposed National Union should be the responsibility of the national body, but that district associations should continue to be responsible for the friendly society function. District associations would fix their own rate of contributions, but they would transmit 5d per member (2½d per half-member) to the national body to finance the Union's industrial work. The National Executive would delegate to district associations responsibility for the settlement of colliery or district questions; although they would reserve the right to exercise overriding authority on all industrial matters. In particular, any dispute likely to lead to a stoppage of work was to be reported to the Executive and no stoppage was to take place without the Executive's sanction.

This draft constitution was considered by the district associations, and various amendments were submitted. In principle however, there was a wide measure of agreement on the form which the reorganized body should take.

The Executive decided to hold a special conference to consider its proposals for reorganization on Wednesday, August 16th, and the succeeding days; and it is not without significance that Nottingham should have been chosen as the meeting-place.

At the outset of the conference, George Spencer, on behalf of his 'Association' welcomed delegates to the City of Nottingham. He was, no doubt, genuinely proud that Nottingham should have been chosen as the scene of this historic meeting, but he lost no opportunity of expressing his opposition to the purpose for which the conference had been called. He opposed proposals coming from South Wales for the Union to adopt as objects 'the complete abolition of capitalism' and the negotiation of 'a National Wages Agreement with the national ascertainment covering the whole of the British Coalfield'. He was defeated on both issues.

He also made much play of the need to protect the position of the officials of the district associations who should be paid a decent wage and who should receive adequate pensions on retirement. Under the draft Rules of the NUM officials were to be compulsorily retired at the age of sixty-five and Spencer suggested that they should be given five years salary as compensation for loss of office. So far as he and his colleagues in Nottinghamshire were concerned, they had service agreements with their organization which provided that they should retain their 'present standard of remuneration'; and he felt that officials in other districts had a moral right to similar treatment. On this issue at least, he had considerable support, and Arthur Horner explained, on behalf of the Executive, that all old officials who were unable to join the Superannuation Scheme on account of age would be given non-contributory pensions.

But on the main point, the formation of one big union, Spencer's was the only voice raised in opposition. He described the proposed constitution as:

'. . . a hybrid conglomeration of nonsense with no uniformity of benefits. Anything short of those things is not one organization at all. You are leaving under the present circumstances a variety of organizations which are different in their functions, different in their intentions, different in the amounts of benefits they are going to pay. Can anyone say that is one organization? No; it is the negation of one organization.

'As you know, I am opposed to one organization, but if there is to be one organization, for heaven's sake let us have one organization.'

He realized all too well that this would be his last conference:

'I am speaking now as a man who is passing out, you can take no notice of me if you like—it does not matter two straws. I am anxious to see this Federation prosper. There is no man more anxious than I am to see this Federation an effective Federation. But I would have liked to keep the local autonomy, honestly I would, with its freedom and all that that means to every one of us.'

The formation of the National Union of Mineworkers was now a foregone conclusion. The debate about the effect of a national ascertainment on the wages of Nottinghamshire miners continued in the local papers and in branch meetings; Spencer criticized the proposal in a leaflet issued to his members, and this brought the Federation's Reorganization Committee to Nottingham to meet the Miners' Council; but the debate left the rank-and-file unmoved and when a ballot was taken in October, an overwhelming majority voted in favour of the formation of the NUM. Even in Nottinghamshire, of 24,001 men voting only 2,836 opposed the proposal. Arrangements were made to register the Rules of the new Union with the Chief Registrar of Friendly Societies as from January 1, 1945; and work was begun on the preparation of Model Rules for the constituent district associations.

V

At the end of 1945 Spencer and Coleman retired, their places as full-time officials being taken by E. J. (Dai) Ley and Harry Straw, both members of the Communist Party, who had the advantage of a well-oiled propaganda machine. William Bayliss was elected to the Presidency of the Area and Herbert Booth became General Secretary.

The Nottinghamshire and District Miners' Federated Union had lasted for little more than seven years before being converted into the National Union of Mineworkers (Nottinghamshire Area). Within the next few years Horace Cooper, William Bayliss and Herbert Booth were to follow Spencer and Coleman into retirement.

The new officials were faced with problems and responsibilities quite different from those of the unhappy years with which this volume has been concerned. Nationalization has, on the whole, led to an improvement in industrial relations; but it has forced upon both sides the need for a new approach to the problems of the industry. Basically however, the task of the Union remains the same; it is to look after the interests of its members; to negotiate price-lists, to settle personal grievances and to provide friendly society benefits and educational facilities. The fact that the Nottinghamshire Area of the NUM does these things rather well is a tribute to the good relations and the organizational forms established by the old union as well as to the way in which the present officials perform their duties.

Many present-day trade unionists appear to believe that the benefits they now enjoy fell like pennies from Heaven. They did not. They had to be planned and fought for. If this volume does nothing else but make that clear, it will have served a useful purpose.

CONCLUSION

The 1920s and 1930s would not, under the most favourable circumstances possible, have been easy years for the miners. The world-wide industrial depression of the 1930s was bound to affect Britain's basic industry at least as much as any other, and more than most. But the actual circumstances were not the most favourable possible. The Government's decision to return to the Gold Standard in 1925 precipitated a slump in Britain whilst the rest of the world enjoyed boom conditions. The coalmining industry, as we saw in Chapter 7, was bound to bear the brunt of the Government's blow to the British economy.

The coal owners were not primarily to blame for the harsh measures of the mid-1920s. Faced with the necessity to slash the price of exported coal coupled with the effect on the home coal market of an artificially induced depression, they had no option but to cut wage rates. But they cannot escape all the blame. Their pusillanimous support for the Government of the day was a matter for regret; their insistence on lengthening the working day was deplorable and their attempt to safeguard profits at the expense of wages was despicable.

The miners were right to oppose any lengthening of the shift; and had the 1926 lockout been fought on this issue it would no doubt, have been successful. In the event the terms which the men were forced to accept at the end of the dispute were worse than the compromise solution which they could have obtained in the early days of the summer whilst the Miners' Federation still retained its strength. The 'Fight to the Finish' advocated by the militants on the left could only have been successful had the Federation been able to starve the economy completely of coal. But they did not possess this power. The Government retained control of the situation throughout. For the vast majority of the miners, the lockout was a simple industrial dispute fought in defence of their living standards. But in order to be successful it would have needed to be converted into a political strike designed to coerce the Government by exercising a stranglehold on the economy. The Federation's Leadership was divided on this issue. People like Varley and Spencer were clearly unwilling to enter the fight at all; whilst the more moderate leaders of the Federation refused to accept the left-wing doctrine of the revolutionary strike. The Secretary, A. J. Cook, combined within himself the schizophrenia of his organization. Emotionally he supported the 'Fight to the Finish' campaign; but intellectually he attempted, however clumsily, to arrange a negotiated settlement on the purely economic issues.

The Nottinghamshire breakaway was a tragedy. The miners of this County have demonstrated over the years their loyalty to their class. But in 1926 circumstances were too much for them. They were driven back to work by hunger and anxiety. Given that the breakaway was inevitable we must nevertheless deplore the subsequent events in the coalfield. The continuance of the division in the miners' ranks was due in part to the intransigence of the coal owners who regarded the Spencer Union as their defence against the militants of the Federation; in part to Spencer's determination to show those who had forced him out of the Federation that he could manage quite well without them; in part to the vested interests of Spencers' supporters; in part to the refusal of many NMA people to seek fusion with the Industrial Union; and finally to the unwillingness of the average Nottinghamshire miner to run the risk of martyrdom.

Fusion, when it came, was forced through by a resurgent Miners' Federation during the boom induced by rearmament. It may be that the terms on which fusion was achieved were dictated by Spencer, but these terms, being based as they were on the substantial autonomy of district associations, were swept into the dustbin of history by the forward movement of the miners which culminated in the formation of the National Union of Mineworkers in 1944 and the nationalization of the industry in 1947.

NOTES

CHAPTER 1

[1] NMA Minutes, 30.1.1915.
[2] ibid., 27.2.1915.
[3] ibid., 27.3.1915.
[4] See Minutes of Conciliation Board. Another 5 per cent advance had been granted in December 1915.
[5] Minutes of the Federated District Conference, June 4, 1918.
[6] On June 30, 1917, Council adopted the motion:
> That District officials' attention be called to the mischievous agita-tion a few men are trying to create in the County, that they be empowered to take any step they think advisable to counteract it and also attend any lodge meeting when requested to do so.'
Presumably this is a reference to the Syndicalist agitation of Jack Lavin of Welbeck and his supporters in the Mansfield district.
[7] NMA Minutes, 16.11.1918.

CHAPTER 2

[1] The Defence of the Realm Act. This Act authorized the making of regulations to ensure the efficient prosecution of the war. The regulations were also used to stifle left-wing opinion.
[2] See speeches of G. A. Spencer at MFGB Special Conference, Southport, 14.1.1919 (Minutes, pp. 37-8); and 15.1.1919 (Minutes, p. 49).
[3] NMA Minute Book, 31.3.1919.
[4] Arbitration Award, 16.6.1920.
[5] See J. E. Williams: 'The Political Activities of a Trade Union', p. 10. (*International Review of Social History*, Vol. II, 1957, Part 1.)
[6] MFGB Minute Book, 1919, p. 437.
[7] NMA Minutes, 28.4.1919.
[8] ibid., 24.2.1919; 28.7.1919; 27.10.1919 & 24.1.1920.

CHAPTER 3

[1] NMA Minutes, 26.4.1920.
[2] MFGB Minute Book, 1920, pp. 807-8. Lloyd George's conditions were not fulfilled: the bottom dropped out of the market.
[3] MFGB Minute Book, 1920, pp. 915-6.
[4] ibid., p. 934.
[5] ibid., p. 937-8
[6] Output in 1919 was 229,779,517 tons compared with 248,499,240 tons in 1917 and 227,748,654 tons in 1918. Output per man-year fell from 248 tons in 1917 to 196 tons in 1919.

[7] Arnot: *The Miners: Years of Struggle*, London, 1952, pp. 231-2.

[8] MFGB Minute Book, 1920, pp. 1375-6. The MFGB asked the NUR to suspend action in view of the negotiations which were to take place with the Government.

[9] Arnot, op. cit., pp. 273-4.

[10] NMA Minutes, 30.10.1920.

[11] *Labour Gazette*, Jan 1921, p. 2, quoted Pigou: *Aspects of British Economic History*, 1918-25, London, 1948, p. 39.

[12] Pigou, op. cit., p. 236.

CHAPTER 4

[1] MFGB Minute Book, 1921, p. 117.

[2] *One Hundred Questions and Answers on Coal* (with a Foreword by Sir Evan Williams, Bart., President of the Mining Association), London, c. 1936, p. 113.

[3] MFGB Minute Book, 1921, p. 116.

[4] Sir Lyndon Macassey: *Labour Policy—False and True*, London, 1922, p. 157.

[5] MFGB Minute Book, 1921, p. 201.

[6] Emphasis supplied.

[7] MFGB Minute Book, 1921, p. 134.

[8] Arnot, op. cit., p. 303.

[9] MFGB Minute Book, 1921, p. 424.

[10] ibid., p. 212.

[11] Arnot, op. cit., p. 331.

[12] Nottingham *Daily Guardian*, July 2, 1921. *The Guardian* correspondent appears to think that wage-earners are the only income losers.

[13] G. C. H. Whitelock: 250 *Years in Coal*, pp. 56-7.

[14] F. Hodges: *Nationalisation of the Mines*, London, 1920, p. 31.

[15] NMA Minute Book, 15.10.1921.

[16] J. W. F. Rowe, *Wages in the Coal Industry*, London, 1923, p. 64.

CHAPTER 6

[1] Dr J. E. Williams tells me that there are references to this body in the Derbyshire *Times*. Mr Herbert Booth has also given me information regarding it. In 1923, the Midland Counties Industrial Protection Society set up by members of the British Workers League had branches at Mansfield, Warsop, Rainworth, Bolsover, Creswell, Rufford and Clipstone.

[2] This same conference took a further important decision: to insist that Frank Hodges should relinquish his post as a full-time official on entering Parliament—MFGB Minute Book, 1923, pp. 700-710.

[3] NMA Minutes, 26.1.1924 & 23.2.1924.

[4] Arnot, op. cit., p. 348.

[5] NMA Minute Book, 29.8.1925.

CHAPTER 7

[1] R. H. Tawney: Article on the Coal Industry in the *Encyclopaedia of the Labour Movement*, Vol. I, pp. 125-6.
[2] Report of Court of Inquiry concerning the Coal Mining Industry Dispute, 1925.
[3] J. M. Keynes: *The Economic Consequences of Mr. Churchill*, p. 6.
[4] Quoted Francis Williams: *Magnificent Journey*, London, 1950, p. 372.
[5] A. Hutt: *British Trade Unionism: A Short History*, London, 1945 Edn., p. 99.
[6] At Ranishaw Park, 2.8.1925. Quoted J. P. Dickie; *The Coal Problem— A Survey*, London, 1936, p. 57.
[7] At Meadowfield, 9.8.1925. Dickie, loc. cit.
[8] W. H. Crook: *The General Strike*, p. 369, quoted Hutt, op. cit., p. 105.
[9] Dickie, op. cit., p. 57.
[10] *Manchester Guardian*, August 4, 1925, quoted Dickie, op. cit., p. 57 and Arnot, op. cit., p. 386.
[11] Report of the Commission, Vol. I, Cmd. 2600, *passim*.

CHAPTER 8

[1] Perhaps the best account of the General Strike so far written is that of Julian Symons: *The General Strike*.
[2] MFGB Minute Book, 1926, p. 339.
[3] ibid., p. 204.
[4] loc. cit.
[5] E. Burns: *The General Strike, May 1926: Trades Councils in Action*; pub. Labour Research Dept., pp. 154-5.
[6] Nottingham *Journal*, May 13, 1926.
[7] However, J. C. C. Davidson (now Lord Davidson) has recently claimed credit for the editorship of the *British Gazette* in a television interview with Julian Symons—See *Readers News*, Jan. 1959.
[8] Samuel Memorandum, para. 4.
[9] ibid., para. 6 (2).
[10] S. Higenbottam: *Our Society's History*, Manchester, 1939, p. 265.
[11] Kingsley Martin: Article on The General Strike in *Encyclopaedia of the Labour Movement*, Vol. II, pp. 26-7.

CHAPTER 9

[1] NMA Minute Book, 24.4.1926; MFGB Minute Book, 1926, pp. 227 & 909.
[2] Dickie, op. cit., p. 75. The speech is summarized in Nottingham *Journal* of 28.5.1926.
[3] MFGB Minute Book, 1926, p. 231.
[4] Nottingham *Journal*, 7.6.1926.
[5] ibid., 13.10.1926.

[6] ibid., 14.10.1926.

[7] ibid., 15.10.1926.

[8] ibid., 20.10.1926.

[9] 'The Trades Union Bill Vindicated by a Labour MP' (a pamphlet containing a report of a Parliamentary speech by G. A. Spencer, MP, published by the Anti-Socialist and Anti-Communist Union) and Nottingham *Journal*, 7.10.1926.

[10] 'Standard' was defined as basis rates, plus the percentage paid on basis rates in 1914, plus any piecework addition for working less than eight hours.

CHAPTER 10

[1] Nottingham *Journal*, 1.11.1926.

[2] ibid., 2.11.1926.

[3] According to Frank Varley—Nottingham *Journal* 3.11.1926. However, according to J. P. Houfton of the Bolsover Company, Varley and his colleagues subsequently asked the owners to meet to discuss a settlement. ibid., 4.11.1926.

CHAPTER 11

[1] This brief account of Wilson's life is based on an unpublished paper by Mr F. Taylor of Ruskin College, Oxford.

[2] See 'The Truth about the Miners Ballot', a reprint of a speech by Walter (now Lord) Citrine at Kirkby on Thursday, May 28, 1928 for details.

[3] Cotter was, prior to 1921, President of a rival union, which was absorbed by Wilson's union in somewhat questionable circumstances. Harry Bond was a Trustee of the NUS.

[4] *Industrial Peace Journal*, July 1927, quoted MFGB Minute Book, 1927-8, p. 1037.

CHAPTER 12

[1] F. Williams, op. cit., p. 400.

[2] MFGB Minute Book, 1927-28, pp. 831-3. The 'understanding' referred to was to have brought to an end the recriminations following the 'betrayal' of the miners by the TUC General Council in 1926.

[3] The voice was the voice of Hicks, but behind it one suspects, was the hand of Citrine.

[4] In Scotland, according to Hutt, the old officials were 'duly and constitutionally voted out by Communists and men of the left', but they refused to vacate their offices—op. cit., p. 117.

[5] From a draft proof prepared for Spencer by the Nottinghamshire coalowners' solicitor in connection with the Coal Mines Nationalization Act, 1946.

CHAPTER 13

[1] In a letter to the author.
[2] In a letter to the author.
[3] A resolution excluding MPS from holding office as Agent was adopted by the NMA on 26.2.1927.
[4] The third member of the sub-committee, Mr Harry Hicken, refused to examine the evidence, and absented himself from most of the meetings.
[5] Whilst the MFGB Officials were conferring at the Miners' Offices, Nottingham on October 19, 1926, a middle-aged gentleman with an attaché-case called to see Cook. He told reporters that he was a party to negotiations between Cook and Winston Churchill and that he had draft terms for a settlement in his case.—Nottingham *Journal* 20.10.1926.

CHAPTER 14

[1] This passage is based on letters from Emmanuel Shinwell and Herbert Booth and talks with the late George Spencer.
[2] Take, for instance, the following extract from the MFGB Minutes (21.5.1931) 'The Committee discussed the position of the Nottinghamshire Miners' Association in relation to the Trades Union position in the county. It was agreed that the officials be empowered to proceed on certain lines with a view to helping the Nottinghamshire Miners' Association.'—MFGB Minute Book, 1931, p. 142.

CHAPTER 15

[1] MFGB Minute Book, 1935, pp. 19 & 43.
[2] Report of Ebby Edwards, Gen. Sec. of the MFGB.—MFGB Minute Book, 1935, p. 13.
[3] Whitelock, op. cit., p. 169.
[4] This account is built up from the MFGB Minute Book, unpublished documents prepared by the owners and the NMIU, a pamphlet by R. Kidd—'The Harworth Colliery Strike'—published by the National Council for Civil Liberties, and talks with old miners.
[5] From a document prepared for the Industrial Union by the owners.
[6] Opinion of Mr D. Bowen in the case of Clifford Parker and the Harworth Colliery Checkweigh Fund Committee.
[7] Nottingham *Journal*, 2.2.1937.
[8] Nottingham *Guardian*, 12.2.1937.
[9] Nottingham *Journal*, 24-26 June, 1937.

CHAPTER 16

[1] MFGB Minute Book, 1939, p. 403 (Report of E. Edwards to Special Conference on October 27, 1939).
[2] The clause in the Resolution to which Spencer referred reads: '(f) Monies allocated from the National Surplus Fund shall be divided as between wages and profit as 90 per cent to wages and 10 per cent to profit.'

BIBLIOGRAPHY

1. North Midland Coalfield; Regional Survey Report, HMSO, 1945.

2. J. U. NEF: *Rise of the British Coal Industry*, London, 1932.

3. *Colliery Year Book and Coal Trades Directory*, London, 1951 Edn.

4. J. E. WILLIAMS: 'The Political Activities of a Trade Union' (*International Review of Social History*, Vol. II, 1957, Pt. 1).

5. G. D. H. COLE and R. POSTGATE: *The Common People* 1746-1938, London, 1938.

6. Coal Industry Commission Reports, Cmd. 359-360 & 361, HMSO, 1919.

7. F. WILLIAMS: *Fifty Years' March: The Rise of the Labour Party*, London, 1950.

8. R. PAGE ARNOT: *The Miners: Years of Struggle*, London, 1952.

9. J. W. F. ROWE: *Wages in the Coal Industry*, London, 1923.

10. A. C. PIGOU: *Aspects of British Economic History*, 1918-25, London, 1948.

11. *One Hundred Questions and Answers on Coal* (with a Foreword by Sir Evan Williams, Bart., President of the Mining Association), London, *circa* 1936.

12. SIR LYNDEN MACASSEY: *Labour Policy, False and True*, London, 1922.

13. G. HARVEY: *Industrial Unionism and the Mining Industry*, Pelaw-on-Tyne, 1917.

14. G. C. D. WHITELOCK: 250 *Years in Coal* (A History of the Barber, Walker Company).

15. F. HODGES: *Nationalisation of the Mines*, London, 1920.

16. H. W. BOOTH: Wages, 1950 (a booklet issued to Branches by NUM Notts. Area).

17. Report of (Buckmaster) Court of Inquiry (Cmd. 2129), HMSO, 1924.

18. R. H. TAWNEY: Article on The Coal Industry in *The Encyclopaedia of the Labour Movement*, Vol. I, London, *circa* 1927.

19. Report of (Macmillan) Court of Inquiry, HMSO, 1925.

20. J. M. KEYNES: *The Economic Consequences of Mr. Churchill*, London, 1925.

21. F. WILLIAMS: *Magnificent Journey*, London, 1954.

22. A. HUTT: *British Trade Unionism: A Short History*, London, 1945 Edn.

23. J. P. DICKIE: *The Coal Problem—A Survey*, 1910-36, London, 1936.

24. *The Coal Crisis: Facts from the Samuel Commission* (Labour Research Dept.), London, 1925.

25. Report of (Samuel) Royal Commission (Cmd. 2600), HMSO, 1925.

26. JULIAN SYMONS: *The General Strike*, London, 1957.

27. EMILE BURNS: *The General Strike, May* 1926: *Trades Councils in Action* (Labour Research Dept.), London, 1926.

28. S. HIGENBOTTAM: *Our Society's History*, Manchester, 1939 (A History of the Amalgamated Society of Woodworkers).

29. KINGSLEY MARTIN: Article on The General Strike in *The Encyclopaedia of the Labour Movement*, Vol. II, London, *circa* 1927.

30. R. PAGE ARNOT: *The General Strike, May 1926: Its Origin and History* (Labour Research Dept.), London, 1926.

31. *The Trades Union Bill Vindicated by a Labour M.P.* (Reprint of a Parliamentary speech by G. A. Spencer, MP in May 1927, issued by the Anti-Socialist and Anti-Communist Union).

32. *The Truth about the Miners' Ballot* (a reprint of a speech by Walter M. (now Lord) Citrine at Kirkby on May 28, 1928).

33. R. KIDD: *The Harworth Colliery Strike* (National Council for Civil Liberties), London, 1937.

34. K. G. J. C. KNOWLES: *Strikes—A study in Industrial Conflict*, Oxford, 1952.

35. Files of the Nottingham *Journal*, Nottingham *Guardian*, Nottingham *Evening Post* and Mansfield *Chronicle-Advertiser*.

36. Minute Books of Nottinghamshire Miners' Association, Nottinghamshire Miners' Industrial Union, Nottinghamshire Miners' Federated Union, Miners' Federation of Great Britain, Federated District Conferences and Eastwood Central Committee.

37. Various papers from collections of G. A. Spencer, DL, JP, O. B. Lewis, and others.

GENERAL INDEX

Abnormal places, 32, 50
Absenteeism, 23, 24, 26, 244, 296
Agricultural Workers' Union 60,
'All-throw-in' system, 53, 54, 55, 61,
   97–100
American War of Independence, 155
Ascertainment, the, 113, 137, 212–19
Annesley, 40, 147, 226, 231, 249, 250, 251

Babbington Collieries, 33, 55, 63
B. A. Collieries, Ltd., 18, 55
'Back to the Union' campaign, 117, 118,
   126, 127, 129, 147
Ballots on N.M.A. *v* Spencer Union
   Issue, 224, 262, 265, 271, 272
Barber, Walker & Co., 18, 52, 59, 96,
   221, 258–73
Basford, 237
Bedworth, 267
Bentinck, 33, 93, 147, 164, 211, 249, 250
Berry Hill Hall Convalescent Homes,
   116
Bentley Colliery, 221
Bestwood, 56, 109, 249
Bilsthorpe, 225, 249, 255
Bircotes, 258
'Bishops Proposals', 165–69
Blidworth, 165, 171
Bolsover, 207, 221
Bolsover Co., 18, 100, 116, 117, 176,
   177, 192
Boys' wages, 32
Breakaway movement in Notts., 174–
   79, 182–202, 203–26, 246, 247, 308
Brickworkers' wages, 53
Brierley Hill, (Sutton Colliery), 33, 60,
   97
Brinsley, 97
Bristol, 94, 124, 125
British Gazette, The, 157, 158
British Worker, The, 157
British Workers' League, 116–17
Broxtowe, 60, 95
Broxtowe Lodge, 278
   (Covering Babbington and Cinderhill
   pits)
Buckmaster Court of Inquiry (1924),
   124
Bulwell, 32, 93, 147

Butterly Co., 18, 59
'Butty" System, 18, 39, 53, 54, 55, 61,
   97–100

Cannock Chase, 91, 112
Central Labour College, 22, 39, 60, 117,
   164, 239
'Chain Gang', The, 262, 263, 289
Checkweighmen, 54, 211, 223, 225, 260
Cheshire, 27
China, 127
Cinderhill, 33, 54, 98, 147
Civil Commissioners, 144, 154
Clerks' pay, 51, 59
Cleveland, 169
Clifton, 249
Clipstone, 165, 171, 176, 190, 192
Coal Charges Fund, 303
Coal Controller, 27, 30, 51, 52, 58, 60, 82
Coal cutters' wages, 32
Cokemen, 91, 145
Collectors, N.M.A., 225, 246, 256, 268,
   281
Communist Party, 39, 115, 142, 159,
   200, 203–5, 233–35, 240, 250, 251,
   304
Cost-of-living, 66, 122, 138, 147, 293,
   298
Cotes Park, 107, 208
Council of Action, 143, 194
Courts of Referees, 248, 249, 276
Crown Farm, 176
Cumberland, 82, 84, 153, 169, 183, 199

Daily Mail, The, 154
Daily Worker, The, 251
Datum line strike, 19, 66, 106
Decontrol of mines, 80–3
Defence of the Realm Acts, 40
Deputies union, 58, 87
Derbyshire, 21, 27, 31, 32, 59, 82, 84,
   86, 89, 91, 92, 98, 99, 111, 112, 118,
   163, 165, 177, 180, 195, 199, 208,
   210, 215, 221, 223, 235, 237, 255,
   256, 265, 294
Digby and New London, 184, 187–89,
   192, 193, 205, 249
Digby Colliery Co., 21, 55, 88, 200–2,
   225

INDEX OF PERSONS

GEORGE ALLEN & UNWIN LTD
London: 40 Museum Street, W.C.1

Auckland: 24 Wyndham Street
Bombay: 15 Graham Road, Ballard Estate, Bombay 1
Buenos Aires: Escritorio 454–459, Florida 165
Calcutta: 17 Chittaranjan Avenue, Calcutta 13
Cape Town: 109 Long Street
Hong Kong: F1/12 Mirador Mansions, Kowloon
Ibadan: P.O. Box 62
Karachi: Karachi Chambers, McLeod Road
Madras: Mohan Mansions, 38c Mount Road, Madras 6
Mexico: Villalongin 32–10, Piso, Mexico 5, D.F.
Nairobi: P.O. Box 12446
New Delhi: 13–14 Asaf Ali Road, New Delhi 1
São Paulo: Avenida 9 de Julho 1138-Ap. 51
Singapore: 36c Prinsep Street, Singapore 7
Sydney, N.S.W.: Bradbury House, 55 York Street
Toronto: 91 Wellington Street West